Sedimentary Processes - Examples from Asia, Turkey and Nigeria

Edited by Gemma Aiello

Published in London, United Kingdom

IntechOpen

Supporting open minds since 2005

Sedimentary Processes - Examples from Asia, Turkey and Nigeria
http://dx.doi.org/10.5772/intechopen.77582
Edited by Gemma Aiello

Contributors
Prince Suka Momta, Takaaki Uda, Shiho Miyahara, Toshiro San-Nami, Masumi Serizawa, Özlem Bulkan, Bilgehan Toksoy, M. Namik Çagatay, Zhonglin Shi, Dongchun Yan, Anbang Wen, Yongyan Wang, Raktima Ghosh, Punarbasu Chaudhuri, Subhamita Chaudhuri, Gemma Aiello

First published in London, United Kingdom, 2020 by IntechOpen
IntechOpen is the global imprint of INTECHOPEN LIMITED, registered in England and Wales, registration number: 11086078, 7th floor, 10 Lower Thames Street, London, EC3R 6AF, United Kingdom
Printed in Croatia

British Library Cataloguing-in-Publication Data
A catalogue record for this book is available from the British Library

Additional hard and PDF copies can be obtained from orders@intechopen.com

Sedimentary Processes - Examples from Asia, Turkey and Nigeria
Edited by Gemma Aiello
p. cm.
Print ISBN 978-1-78984-764-2
Online ISBN 978-1-78984-765-9
eBook (PDF) ISBN 978-1-83880-729-0

For EU product safety concerns:
IN TECH d.o.o., Prolaz Marije Krucifikse Kozulić 3, 51000 Rijeka, Croatia, info@intechopen.com or visit our website at intechopen.com.

Meet the editor

Dr. Gemma Aiello was born in Aversa (CE), Italy, on 24 October 1964. In 1989, she graduated in Geological Sciences at the University of Naples "Federico II". In 1993, she earned a PhD degree in Sedimentary Geology at the University of Naples "Federico II", Department of Earth Sciences, Faculty of Geological Sciences. She completed a 2-year postdoctoral fellowship at the University of Naples "Federico II", a CNR-CEE fellowship, and several contracts at the Research Institute "Geomare Sud", CNR, Naples, Italy. Since 1998, she has been a full-time researcher at the Italian CNR. Dr. Aiello has 25 years' experience in the field of sedimentary geology, marine geology, and geophysics, participating in different research projects for the Italian National Research Council (CARG, Vector, Centri Regionali di Competenza). She was a contract professor of sedimentology and stratigraphy at the Parthenope University of Naples, Italy, and a teacher in formation courses of technicians in marine science and engineering in Naples, Italy.

Contents

Preface

In this book, “Sedimentary Processes – Examples from Asia, Turkey, and Nigeria”, the geological characteristics of different depositional environments have been singled out, particularly referring to deltas, beach systems, and coastal lacustrine environments. In this book, the other research topics are the riparian zone and the role of mangroves in controlling coastal sedimentation. The sedimentary processes represent one of the most important control factors influencing the facies distribution in a stratigraphic succession and are critical in coastal environments. In deltaic systems the vertical arrangement of depositional cycles is mainly controlled by the decrease of the gradients due to the progradation of the distributary channels. Exceptional events, including floods, storms, and earthquakes also influence the corresponding changes in the sedimentary patterns. The sedimentary processes are strictly related to the role of the sedimentary supply in controlling the stratigraphic architecture of the sedimentary sequences, depending on the availability of sediments, subsidence, and relative sea-level changes. In this framework, two situations may be distinguished, i.e. transgressive and regressive. In the transgressive situation, the subsidence and the sea-level rise are more important than the terrigenous supply and a sediment starvation occurs. Reworking, erosion, and diagenesis take place, coupled with an increase of chemically and biologically formed sediments. In the regressive situation, the subsidence and the sea level rise are less important than the terrigenous supply, resulting in progradation and increase of continental facies. Another important control factor acting on sedimentary processes is represented by the climate, due to the climate affecting the facies apart from temperature and rainfall, seasonal extremes, and sporadic fluctuations. Relative sea-level changes have been brought about by the sediment progradation, by the vertical movements, or by the tectonic tilting of the crustal blocks, by the changes in volume of the oceanic waters, or by the global tectonic changes, such as the variation in volume of the oceanic ridges. Delta models have individuated high constructive (tide-influenced and wave-influenced) deltas and high destructive (lobate and elongate) deltas. Each delta type is distinguished from a characteristic morphology and facies pattern, which has been described including the vertical sequences, the facies associations, the facies distribution, and the geometry of the sandy bodies and reservoirs. The delta progradation produces a typical facies succession, consisting of prodelta facies, delta front, and delta plain facies. A coarsening upward succession of prodelta sediments to delta front sediments is overlain by a fining upward succession, composed of upper delta front sediments to delta plain sediments. The lacustrine deposits have a high potential of preservation and perhaps they represent excellent sedimentary archives in palaeo-climatic reconstructions. The sedimentary records of varves, annually laminated, provide high-resolution archives of palaeo-environmental conditions, so providing both chronological and geochemical information. Beach depositional systems and their changing morphology have been controlled by storm events, changing the coastal morphology, and triggering the formation of megacusp embayments. The impact of storms on the variation of the beach has been analyzed by combining the historical measurements of the beach profile surveys and the numerical modeling of storm-induced beach changes. The role of mangroves in controlling the coastal sedimentation is critical, since they allocate more carbon below the ground, representing rich-carbon biomes, which facilitate the deposition of fine-grained sediments.

This book, made up of six chapters, examines different studies on the sedimentary processes, including the geologic characteristics. The first chapter introduces the sedimentary processes and related geological controls, examining the role of the deltas (Niger Delta), the coastal and lacustrine geo-archives, the beach systems and their morphological variations, and the geological role of mangroves. Using geophysical techniques for the analysis of the seismic attributes, in the second chapter it has been established that the N5.2 reservoir is a massive sandy unit, located in the Agbada Formation, representing a structural high, characterized by two main peaks. The correlation of well logs was performed, highlighting the occurrence of sandstones with traces of glauconites. In the third chapter, the coastal and lacustrine sedimentary archive of Lake Bafa was analyzed, individuating four main depositional stages, ranging in age from 4.5 ky B.P. to 0.8 ky B.P. The ecosystem characteristics of the basin have been controlled by hydroclimatic and geotectonic processes. The fourth chapter applied a fingerprinting technique to the sediment sources of the riparian zone, taking into account several depths of a tributary channel. The main sediment sources are represented by the topsoil above the riparian zone, by the sediments suspended by the Yangtze river, and by the sediment suspended by the Ruxi river. The fifth chapter has studied the long-term variations of the beach of the Kaike coast (Japan), starting from the analysis of aerial photographs and then producing a contour-line change model, which has considered the variations of sediment grain-size. In the sixth chapter, marker horizon techniques and surface elevation table (SET) tests have highlighted the spatial variability in the sedimentary patterns by examining several coastal sites, located in Malaysia and Thailand (south-eastern Asia).

I thank Mrs. Romina Skomersic, who has followed the publication process with competence and patience, allowing for the development of this book process.

Dr. Gemma Aiello, PhD
Istituto di Scienze Marine (ISMAR),
Consiglio Nazionale delle Ricerche (CNR),
Sezione Secondaria di Napoli,
Napoli, Italy

Chapter 1

Introductory Chapter: An Introduction to Sedimentary Processes - Examples from Asia, Turkey, and Nigeria

Gemma Aiello

1. Introduction

This is the introductory chapter of the book *Sedimentary Processes - Examples from Asia, Turkey and Nigeria*. In this book, different topics on the sedimentary processes have been treated, focusing, in particular, on the geological characteristics and production response of a sandstone reservoir located offshore the Niger Delta (Nigeria); on coastal lacustrine sedimentary archives with the example of Lake Bafa (Mediterranean Sea); on the riparian zone of the Three Gorges Reservoir (China), representing a transitional area located between aquatic and terrestrial environments; on the long-lasting changes of the beach environments along the Kaike coast (Japan); and, finally, on the role of the mangroves in coastal and estuarine sedimentary environments.

The facies distribution and the corresponding changes are strongly influenced by several control factors, including the sedimentary processes, the sediment supply, the climate, the tectonics, the sea level changes, the biological activity, the water chemistry, and the volcanism. In different depositional environments, these control factors are of variable importance, but the climate and the structural setting act on the whole sedimentary environments. On the other side, the sedimentary processes are critical in deltaic and fluvial environments. On the continental margins, relative sea level fluctuations involve the shallow seas and the shorelines, more than in the continental and deep marine environments, also if their effects are not negligible [1–4]. In a given depositional environment, the sedimentary processes, represented by the processes intrinsic to sedimentation, are responsible for the facies distribution and change. For instance, the progradation of the distributary channels of a delta controls a decrease of the gradient, so that the river has a short route, starting a new depositional cycle. Due to the nature itself of the depositional environments, these kinds of changes are inevitable, since the timing of these changes is controlled by unusual events, such as floods, storms, or earthquakes. These trigger causes must be distinguished from the fundamental causes represented by the delta progradation [5–7], by the river aggradation, and by the slope instability [8–10]. Sediment supply represents another important control factor. Its effect depends on the sediment availability, subsidence, and relative sea level changes. In this book, the main research topics are represented by the deltas, particularly referring to the Niger Delta; by the coastal and lacustrine sedimentary archives, particularly referring to the Lake Bafa; by the riparian zone; by the beach environments and their variations during the geological time; and, finally, by the

IntechOpen

role of the mangroves in the coastal sedimentation. These topics will be outlined in the following sections.

2. The role of deltas in the sedimentary processes and the Niger Delta

The deltas are protuberances of the shoreline, which were formed when the rivers enter the oceans, the semi-enclosed seas, the lakes, or the lagoons sheltered by barriers. They rapidly supply sediments, which can be redistributed by basinal processes. These deltas are served by drainage systems, which are well developed and culminate in a trunk stream, supplying sediments to a restricted area of the shoreline. The drainage systems produce closely spaced rivers, inducing a uniform progradation of the whole coastal plain rather than a point-concentrated progradation. The studies of the deltaic facies started in ancient successions rather than in the modern deltas with the Gilbert deltas [11, 12], describing the Pleistocene deltaic facies in Lake Bonneville. The delta has a three-dimensional structure, generating a distinctive vertical sequence of types of bedding during the delta progradation. From the lower part of the sequence to the upper part of the sequence, there is the bottomset, composed of gently inclined fine-grained sediments; the foreset, composed of beds of sands and gravels dipping from 10 to 25°; and the topset, which is composed of flat-lying gravels [11, 13]. These terms (topset, foreset, and bottomset) have been used in order to describe the delta structure and the bedding, texture, color, and fauna of each component. Although not all the deltas show a Gilbert-type structure, these concepts have conditioned the thinking on modern deltas for several decades. Therefore, the occurrence or the lack of large and inclined foresets has been considered as an important criterion for the study of deltaic successions. Moreover, the economic relevance of deltaic facies has stimulated the execution of wide borehole programs in the Mississippi, Rhone, and Niger deltas [14–19]. These studies have shown that the deltaic successions include a variety of vertical facies and sequences and that a type of sequence within a delta varies at different locations, as well as within the deltas. The conceptual framework of the comparative studies on deltas starts from the hinterland characteristics, controlling the fluvial regime and the sediment input, which influence the delta regime (controlled also by the basinal regime), the delta morphology, and the delta facies pattern. Of course, the delta types have been defined based on the depositional regime and illustrated by a characteristic morphology.

The Niger Delta is one of the most important research topics of this book. It is a large, arched delta of a destructive wave-dominated type. A succession of marine clays, overlain by paralic deposits, in turn covered by continental sands, occurs. This sequence has been built up by superimposed offlap cycles. Basement faulting has affected the development of delta and, consequently, the sediment thickness distribution. In the paralic interval, growth fault-associated rollover structures have trapped hydrocarbons. For this reason, the Niger Delta hosts main hydrocarbon reservoirs. In this case, the growth faults have functioned as hydrocarbon migration pathways from the overpressured marine clays [20].

3. The coastal lacustrine sedimentary archives and the Lake Bafa

The coastal lacustrine sedimentary archives have been deeply studied, particularly referring to their use in paleoclimatic reconstructions and to the lithological and geochemical aspects [21–24]. Zolitschka [21] has outlined that the lacustrine sediments have a high potential as proxies in paleoclimatic reconstructions. The annually laminated (varved) sedimentary records in a lacustrine environment represent important high-resolution archives of paleoenvironmental conditions. The most important control factor is represented by the climate and by the anthropogenic change. The linkage between the climate and the varves has been deeply studied, in particular for the proglacial lakes linked with the clastic varves (mean summer temperature, mean summer precipitation). At the middle latitudes, the varves are also controlled by the organic productivity. Moreover, the thickness of the varves may increase with the minerogenic detritus, which can be regarded as a discharge proxy. Two kinds of information have been provided, chronological and geochemical. The varve chronology has to be calibrated through other and more precise dating methods due to several misinterpretations (Holocene sediments of Skilak Lake, Alaska; Precambrian laminites, Australia). The geochemical composition of the yearly element has been calculated. The geochemical analyses have been performed on samples having a thickness of at least 1 cm, including several years or decades of deposition. Baroni et al. [22] have analyzed a core retrieved from the Lake Frassino (northern Italy), which has provided evidence of main paleohydrological changes during the last 14 ky B.P. The lake evolution has been reconstructed during the Late Glacial and the Holocene by using lithological, malacological, and isotopic composition of freshwater shells. During the Late Glacial, the conditions were drier than in the Holocene, and a wetter period has been suggested to occur before 14 ky B.P. The oxygen isotopic data have suggested a clear bipartition during Holocene times, with a dry first part (9100–7000 years B.P.), followed by an increase in humidity (7000–6800 years B.P.), while from 5000 to 2600 years, the isotopic record was characterized by large fluctuations, suggesting alternating wet and dry periods. Basilici [23] has studied the lacustrine facies of the Tiberino Basin, which was formed after Plio-Pleistocene tectonic phases of the central Italy. Four facies associations have been distinguished, consisting of the facies association A, which was deposited in a deep-offshore lacustrine environment, consisting of massive, laminated, bluish-gray marly clays and representing the main lacustrine deposit. The other facies associations represent the marginal facies. The facies association B has been interpreted as a Gilbert-type delta system, showing gravel bodies and prodelta bodies, consisting of marly clays, alternating with sands and gravelly mud strata. The facies association C corresponds to the coastal environment, composed of interbedded muddy and sandy strata and clayey silts and lignites, interpreted as a coastal wetland. The facies association D is represented by the distal part of a muddy alluvial fan. The paleoenvironmental reconstruction has indicated that the Tiberino Basin hosted a narrow lake during Pliocene times, whose size, shape, and depth were controlled by the tectonic setting. Beck [24] has highlighted that the Late Quaternary sedimentary fill of several lakes, located in the northwestern Alps, may represent a paleoseismological sedimentary archive. These peculiar strata have been controlled by mass failures or subaqueous slope deposits (delta foresets), evolving into hyperpicnal currents, and by in situ liquefaction and flowage, more than by micro-fracturing. The paleoseismic interpretation has been extrapolated up to 16 ky B.P., reconstructing the time series and identifying from a textural point of view several kinds of slope failure deposits. This has allowed to obtain the temporal series, which are compatible with the historical seismicity due to the observed recurrence interval.

The Lake Bafa is a saline-brackish wetland ecosystem, having an international importance, which is located at the southeastern part of Büyük Menderes River Delta [25]. It is bounded by the Beşparmak Mountains to the south and to the east and by the alluvial plains of Büyük Menderes Delta to the north and to the west [25]. The lake's surface area is 6708 hectares for 25 m depth, while the average water level is only 5 m [25]. The main water source is represented by the Büyük Menderes River, but some small streams have also contributed to the water input in the lake.

The Lake Bafa has been formed as the result of delta progradation of the Büyük Menderes River. The sediments of the river have filled the marine embayment (Latimian Gulf). During the last millennia, the Latimian Gulf has been transformed into a deltaic and alluvial plain.

4. The beach environment and its variation during geological time

The beaches and the barrier islands are long and narrow sand accumulations, occurring within the deltas, along the depositional strike from deltas or in an oceanic or lacustrine context, without any relationship with the deltas. Both the depositional systems are aligned parallel to the shoreline. The beaches are attached to the land, while the barrier islands are separated from the land by a shallow lagoon and are often dissected by tidal inlets. The formation of the beaches and of the barrier island systems includes a steady supply of sands to the shoreline (river input, longshore drift) and a hydrodynamic setting characterized by low and moderate wave energy but a limited tidal range. The beaches and the barrier islands have been constructed by the wave processes, which have been intensively studied through direct observation, experimentation, and theoretical procedures. Regarding the wave processes controlling the beaches, the first process to be discussed is the wave transformation, the second one is represented by the wave-induced nearshore currents, and the third one consists of the temporal variations in the wave regime. The major storm events have the role to control the landward retreatment or the local breaching of the eolian dune ridge. The beach includes a variety of sub-environments, including the eolian sand dunes, the backshore-foreshore, and the shoreface. In particular, the eolian sand dunes form complex ridges, capping the beach face above the mean tide level and resulting from wind reworking of sands emplaced in the upper beach by the storm waves, attaining a height of several meters. The backshore represents the supratidal part of the beach, which is flooded during the storm events, whereas the foreshore represents the intertidal area. The shoreface is the subtidal part of the beach, starting at the mean low tide level and terminating at the fair-weather wave base.

5. The role of mangroves in coastal sedimentation

In the coastal protection, the root systems of the mangrove forests trap sediments flowing down rivers and towards the land. This allows to stabilize the coastline and prevents the erosion operated from waves and storms. In the areas where the mangroves have been cleared, the coastal damage from hurricanes and typhoons is stronger. The role of mangroves in the coastal sedimentation has been deeply studied [26–43]. Alongi [26] has studied the carbon sequestration in the mangrove forests. The mangrove forests are highly productive, with the carbon production rates which are equivalent to the tropical humid forests. The mangroves host more carbon below the ground and have higher carbon mass ratios than the terrestrial trees. The most of the mangrove carbon is stored as large pools in the soil and dead roots. Moreover, the mangroves account for only approximately 1% of carbon sequestration by the world's forests, but as coastal habitats they account for 14% of carbon sequestration by the global ocean. Banerjee et al. [27] have studied four sediment cores from selected locations of the Sundarbans mangroves and Hooghly estuary (northeastern coast of India) to reconstruct the ^{210}Pb geochronology and to individuate the trace metal distribution in the sediments. The mangroves of India account for about 5% of the world's mangrove vegetation and are spread over an area of about 6740 km^2. The region of Sundarbans hosts the 10% of the mangrove forests in the world and a half of the total area under mangroves in India. The Sundarbans mangroves and the Hooghly estuary have received a considerable pollution load from anthropogenic sources such as the industrial, the domestic, and the shipping activities in recent times, suggesting a high concentration of metals in the top few layers. The obtained results have suggested that the variation in trace metal content with depth or between mangrove and estuarine systems derives from the metal input due to the anthropogenic activities rather than to the diagenetic processes. Blasco et al. [28] have examined the mangroves as indicators of coastal changes, studying in which way these ecosystems have been used as indicators of coastal changes or sea level rises. These ecosystems are specialized, and any minor variation of their hydrological and tidal regime controls a noticeable mortality. The mangroves are highly sensitive to the inundation regime. If the tectonic, sedimentologic, and hydrological events have been modified, these species tend to readjust to new environmental conditions, or alternatively, they tend to succumb to unsuitable environmental conditions. Perhaps, the use of the remote sensing data for the mangrove ecosystem represents a good tool in the coastal monitoring. In this book, the role of mangroves in the coastal and estuarine sedimentary accretion of southeastern Asia has been discussed. In fact, the mangroves provide also characteristic mechanisms in order to trap the sediments and to accelerate the land-building processes in the tide-dominated coastal and estuarine environments.

6. Outline

This topic examines different studies on the sedimentary processes, including:

a. The geologic characteristics and the production response of the N5.2 reservoir, located offshore the Niger Delta and evaluating the geological elements, mainly the sedimentary facies and the structural lineaments, which have controlled the decline in the reservoir production

b. The example of the Lafe Bafa as an excellent geo-archive located in the Mediterranean sea, studied through lithostratigraphy, chronostratigraphy, and geochemical data, providing evidence for a continuous accumulation during the last 4.5 ky

c. The sedimentary processes in the riparian zone of the Ruxi Tributary Channel (Three Gorges Reservoir, China) applying a composite fingerprinting technique to apportion the sediment sources for the riparian zone with different elevations and studying the sedimentary input from this channel as a main source of pollution for the riparian environment

d. The long-term changes in the beach environments along the Kaike coast (Japan), reproduced using a contour-line change model which has taken into account the grain size of the beach sediments and evaluating the long-shore transport of sands through bathymetric data analysis

e. The role of the mangroves in the coastal and estuarine sedimentary processes in southeastern Asia, highlighting the sediment accretion between the different types of roots, the spatial variability of the sediment accretion, and the influence of the seasonal change impacts on the sediment accretion

The relationships of the sedimentary processes with the sea level changes and the subsidence have also been examined.

Author details

Gemma Aiello
Institute of Marine Sciences (ISMAR), National Research Council of Italy (CNR), Naples, Italy

*Address all correspondence to: gemma.aiello@iamc.cnr.it; gemma.aiello@na.ismar.cnr.it

References

[1] Haq BU, Hardenbol J, Vail PR. Chronology of fluctuating sea levels since the Triassic. Science. 1987;**235**(4793):1156-1167. DOI: 10.1126/science.235.4793.1156

[2] Blum MD, Tornquist TE. Fluvial responses to climate and sea-level change: A review and look forward. Sedimentology. 2000;**47**:2-48

[3] Catuneanu O, Abreu V, Bhattacharya P, Blum MD, Darlymple W, Eriksson PG, et al. Towards the standardization of sequence stratigraphy. Earth-Science Reviews. 2009;**92**(1-2):1-33

[4] Catuneanu O, Galloway WE, St KCG, Miall AD, Posamentier HW, Strasser A, et al. Sequence stratigraphy: Methodology and nomenclature. Newsletters on Stratigraphy. 2011;**44**(3):173-245

[5] Ninfo A, Ciavola P, Billi P. The Po delta is restarting progradation: Geomorphological evolution based on a 47-years earth observation dataset. Scientific Reports. 2018;**8**:3457

[6] Maselli V, Trincardi F. Man-made deltas. Scientific Reports. 2013;**3**:1926

[7] Correggiari A, Cattaneo A, Trincardi F. The modern Po delta system: Lobe switching and asymmetric prodelta growth. Marine Geology. 2005;**222/223**:49-74

[8] Sultan N, Cochonat P, Canals M, Cattaneo A, Dennielou B, Haflidason H, et al. Triggering mechanisms of slope instability processes and sediment failures on continental margins: A geotechnical approach. Marine Geology. 2004;**213**:291-321

[9] Masson DG, Wynn RB, Talling PJ. Large landslides on passive continental margins: Processes, hypotheses and outstanding questions. In: Mosher DC et al., editors. Submarine Mass Movements and their Consequences. Advances in Natural and Technological Hazards Research. Vol. 28. Dordrecht: Springer; 2010

[10] Leynaud D, Mienert J, Vanneste M. Submarine mass movements on glaciated and non-glaciated European continental margins: A review of triggering mechanisms and preconditions to failure. Marine and Petroleum Geology. 2009;**26**:618-632

[11] Gilbert GK. The topographic features of lake shores. Annual report of the United States Geological Survey. 1885;**5**:69-123

[12] Gilbert CK. Lake Bonneville. U.S. Geological Survey. 1890;**1**:1-438

[13] Barrell J. Criteria for the recognition of ancient delta deposits. Bulletin Geological Society of America. 1912;**23**:377-466

[14] Fisk HN, Mc Farlan E Jr, Kolb CR, Wilbert LJ. Sedimentary framework of the modern Mississippi delta. Journal of Sedimentary Petrology. 1954;**24**:76-99

[15] Fisk HN. Sand facies of recent Mississippi delta deposits. In: World Petroleum Congress. Tulsa, Oklahoma: Rome; 1955. pp. 377-398

[16] Fisk HN. Bar finger sands of the Mississippi delta. In: Peterson JA, Osmond JC, editors. Geometry of Sandstone Bodies – A Symposium. Tulsa, Oklahoma: The American Association of Petroleum Geologists; 1961. pp. 29-52

[17] Oomkens E. Depositional sequences and sand distribution in a deltaic complex. Geologic en Mijnbouw. 1967;**46**:265-278

[18] Oomkens E. Lithofacies relations in late quaternary Niger delta complex. Sedimentology. 1974;**21**:195-222

[19] Weber KJ. Sedimentological aspects of oilfields of the Niger delta. Geologic en Mijnbouw. 1971;**50**:559-576

[20] Daukoru J. Petroleum geology of the Niger Delta. World petroleum congress. Oklahoma: In 9th World Petroleum Congress. Tokyo, Japan; 1975

[21] Zolitschka B. High resolution lacustrine sediments and their potential for palaeoclimatic reconstruction. In: Jones PA, Bradley RS, Jouzel J, editors. Climatic Variations and Forcing Mechanisms of the Last 2000 Years. Berlin Heidelberg: Springer-Verlag; 1996

[22] Baroni C, Zanchetta G, Fallick AE, Longinelli A. Mollusca stable isotope record of a core from Lake Frassino, northern Italy: Hydrological and climatic changes during the last 14 ka. The Holocene. 2006;**16**(6):827-837

[23] Basilici G. Sedimentary facies in an extensional and deep-lacustrine depositional system: The Pliocene Tiberino Basin, Central Italy. Sedimentary Geology. 1997;**109**:73-94

[24] Beck C. Late quaternary lacustrine paleo-seismic archives in North-Western Alps: Examples of earthquake-origin assessment of sedimentary disturbances. Earth-Science Reviews. 2009;**96**:327-344

[25] Geography of Lake Bafa. www.lakebafa.org. Contact Tel: 0531 5603018

[26] Alongi DM. Carbon sequestration in mangrove forests. Carbon Management. 2012;**3**(3):313-322

[27] Banerjee K, Purvaja R, Senthilkumar B, Ramesh R. Sedimentation and trace metal distribution in selected locations of Sundarbans mangroves and Hooghly estuary, northeast coast of India. Environmental Geochemistry and Health. 2012;**34**:27-42

[28] Blasco F, Saenger P, Janodet E. Mangroves as indicators of coastal change. Catena. Tulsa, Oklahoma. 1996;**27**:167-178

[29] Woodroffe C. Mangrove sediments and geomorphology. In: Bowman MJ, Barber RT, CNK M, Raven JA, editors. Coastal and Estuarine Studies. Vol. 41. Tulsa, Oklahoma: American Geophysical Union; 1992. pp. 7-42

[30] Duarte CM, Losada IJ, Hendriks IE, Mazarrasa I, Marbà N. The role of coastal plant communities for climate change mitigation and adaptation. Nature Climate Change. 2013;**3**:961-968

[31] Ellison JC. Impacts of sediment burial on mangroves. Marine Pollution Bulletin. 1998;**37**(8-12):420-426

[32] Furukawa K, Wolanski E. Sedimentation in mangrove forests. Mangroves and Salt Marshes. 1996;**1**(1):3-10

[33] Furukawa K, Wolanski E, Mueller H. Currents and sediment transport in mangrove forests. Estuarine, Coastal and Shelf Science. 1997;**44**:301-310

[34] Gedan KB, Kirwan ML, Wolanski E, Barbier EB, Silliman BR. The present and future role of coastal wetland vegetation in protecting shorelines: Answering recent challenges to the paradigm. Climatic Change. 2011;**106**:7-29

[35] Hait AK, Behling H. Holocene mangrove and coastal environmental changes in the western Ganga–Brahmaputra Delta, India. Vegetation History and Archaeobotany. 2009;**18**:159-169

[36] Holmer M, Olsen AB. Role of decomposition of mangrove and seagrass detritus in sediment carbon and nitrogen cycling in a tropical mangrove forest. Marine Ecology Progress Series. 2002;**230**:87-101

[37] Lee SY, Primavera J, Dahdouh-Guebas F, McKee K, Bosire JO, Cannicci S, et al. Ecological role and services of tropical mangrove ecosystems: A reassessment. Global Ecology and Biogeography. 2014;**23**:726-743

[38] Lovelock CE, Feller IC, Ellis J, Schwarz A, Hancock N, Nichols P, et al. Mangrove growth in New Zealand estuaries: The role of nutrient enrichment at sites with contrasting rates of sedimentation. Oecologia. 2007;**153**:633-641

[39] Marchand C, Lallier-Verges E, Baltzer F, Alberic P, Cossa D, Baillif P. Heavy metals distribution in mangrove sediments along the mobile coastline of French Guiana. Marine Chemistry. 2006;**98**:1-17

[40] Rivera-Monroy VH, Twilley RR. The relative role of denitrification and immobilization in thefate of inorganic nitrogen in mangrove sediments (Terminos lagoon, Mexico). Limnology and Oceanography. 1996;**41**(2):284-296

[41] Sahoo K, Dhal NK. Potential microbial diversity in mangrove ecosystems: A review. Indian Journal of Marine Sciences. 2009;**38**(2):249-256

[42] Spalding MD, Ruffo S, Lacambra C, Meliane I, Shepard CC, Beck MW. The role of ecosystems in coastal protection: Adapting to climate change and coastal hazards. Ocean and Coastal Management. 2014;**90**:50-57

[43] Tue NT, Ngoc NT, Quy TD, Hamaoka H, Nhuan MT, Omor K. A cross-system analysis of sedimentary organic carbon in the mangrove ecosystems of Xuan Thuy National Park, Vietnam. Journal of Sea Research. 2012;**67**:69-76

Chapter 2

Geologic Characteristics and Production Response of the N5.2 Reservoir, Shallow Offshore Niger Delta, Nigeria

Prince Suka Momta

Abstract

Sedimentary facies and structural lineaments represent significant control factors on hydrocarbon flow behavior. These geological elements have been evaluated to understand possible cause(s) of rapid decline in pro¬duction. The N5.2 reservoir, located in shallow marine sandstones, offshore Niger Delta, has experienced decline in oil rate with a corresponding increase in water-cut within two years of beginning of production. The main objective of this study is the determination of reservoir architecture in order to individuate the possible cause(s) of rapid production decline. To this aim, several methods have been used, including the seismic attribute analysis, electrofacies analysis, well log and petrophysical correlations. The obtained results show that the N5.2 reservoir is a massive sandy unit, occurring within the paralic Agbada Formation of about 2133 m thick. A contour depth map of the reservoir shows the occurrence of a structural saddle associated with an elongated closure having two structural culminations. Further analysis using the root mean square (RMS) and anti-tracking seismic attributes has indicated a seismic facies parallel to the paleo-coastline direction and several faults and fractures. The high quality of the reservoir, fractures, poor management and water injection may have induced rapid fluid flow and consequently early watercut and decline in production.

Keywords: sedimentation, oil rate, watercut, porosity, permeability, reservoir

1. Introduction

Decline in oil production is common to mature oilfields where several wells have been drilled and exploited over a period of time. The volume of oil produced and the lifetime of a specific well or reservoir is a function of several variables, which could include the thickness of the oil column, sand-body architecture and geometry, quality of hydrocarbon, and several geologic uncertainties. It is expected that after many years of oil production from a reservoir there should be a time when the water oil ratio (WOR) should be significant. However, early decline in production for a reservoir with high potential is uncommon. The studied reservoir has experienced rapid decline in oil production with a corresponding rapid water breakthrough in just 2 years of the beginning of production.

Stratigraphic features and structural lineaments are geologic elements that will be examined to understand the architecture of the reservoir and the possible effects of lineaments on fluid flow. Fluvial and marine processes have influenced the deposition of sediments in the Niger Delta [1, 2]. Myriads of sub-environments exist within each mega environment either continental or marine. The stratigraphic imprint of these events at various scales can be identified and reconstructed from sedimentary rock bodies in both outcrops and cores [3, 4].

The structural elements that influence fluid flow include fractures, faults and folds, mapped in the subsurface of the Niger Delta [1, 5]. They are significant for the hydrocarbon entrapment and fluid mobility within the reservoir. Faults play an important role in the distribution of fluids in the subsurface and may act as baffles or conduits to flow [6]. In particular, this paper focuses on stratigraphic and structural features and their impact on the hydrocarbon production.

It has been established that facies architecture and reservoir condition influence the movement of fluids within the subsurface [4, 7–9]. However, a few studies have examined the control of geological factors in combination with production management strategy on flow behavior [7, 9, 10]. An integrated approach is adopted in this study to examine the relative impact of a combination of these uncertainties on fluid flow especially water, that can result in a rapid production decline. Rock types and fabrics are more significant than fluid properties in controlling the stratigraphic architecture of the reservoir. This is because, geologic characteristics such as facies properties and architecture, changes in depositional environments, etc., have more control on oil and gas recovery than production methods [11, 12]. In the Niger Delta, most of the sand-bodies hosting hydrocarbon occur as channels, shoreface, beach and barrier sand deposits [2, 13–15].

Studies on reservoir characterization of the Niger Delta have been carried out by several researchers. Osinowo et al. [16] described and characterized the Eni field, located offshore Niger Delta by integrating both seismic and wireline logs. Oyedele et al. [17] discussed structural lineaments and their impact on hydrocarbon accumulation in the EMI field, Niger Delta. The work of [18] focused on the impact of facies on reservoir quality and the application of 3D static model to characterize hydrocarbon potential of the KN field offshore, Niger Delta. The interplay of marine energy fluxes, the creation of accommodation space, sedimentation, facies architecture and hydrocarbon potentials of deep offshore areas have been explained in articles by [4, 19]. Reservoir characterization is essential at every stage in the lifetime of an oilfield for appropriate description of sand-body geometry and architecture [20–23].

This study aims at establishing facies characteristics, lateral continuity of sand bodies and the possible cause(s) of early water breakthrough that result in rapid decline in oil production of the N5.2 sand. This concept will have broad significance in the following areas: risk assessment and drilling optimization, understanding the geometries of sedimentary rocks and changes in their associated porosity and permeability [20].

2. Study area

The study area lies in the shallow offshore area of the Niger Delta, Nigeria (**Figure 1**). It is bounded in the north by Calabar Town, at a distance of about 3 km, and in the east by the Cameroon Volcanic Line about 30 km away. The field is circa 229 km^2, and the structure was described by [24], as an elongate, four-way dip closure in two culminations with a saddle between them. There are over 40 wells drilled in the field some of which include pilot holes, water injectors and horizontal production wells.

Figure 1.
Map showing the study location.

3. Reservoir stratigraphic architecture, facies and well log analysis

The field covered a distance of more than 7 km from the east to the west. The lithologic well correlation indicated that most of the sand-bodies (N5.2 and others) are laterally extensive across the oilfield (**Figures 2** and **3b**). A key stratigraphic marker used to constrain the correlation is the Qua Iboe Shale (QIS) (**Figure 3a**). Gamma ray trends forms the basis for delimiting reservoir tops and bases (**Figure 3a** and **b**).

The N5.2 sand is a high quality reservoir characterized by sand with some silt and shale fractions in almost negligible proportions. It is laterally extensive and cuts across the field covering a distance of more than 7 km, and well developed towards the eastern flank. It displays a layer-cake architecture with gross thickness between 33.82 and 105.13 m. The N5 reservoir displays stacked highstand parasequence set and occurred in six sub-units in wells located in the eastern part of the field (**Figure 3a** and **b**).

3.1 Environment of deposition

The well log (gamma ray log) and drill-cutting samples are the major lithologic tools used to identify lithofacies in the field. Ditch cutting sample description (**Table 1**) [15] indicates that the N5.2 sand contains some glauconite pellets showing sedimentation within a marine environment. The depositional environments inferred using electrofacies and the presence of glauconite within the sample interval are; beach, barrier bar, shoreface and regressive bars [15]. The overall gamma ray trend shows an upward coarsening sequence.

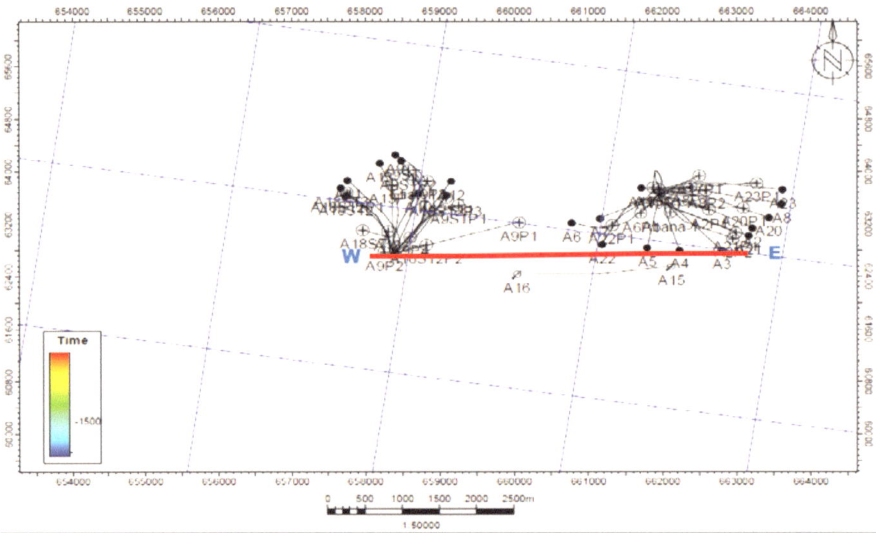

Figure 2.
Seismic base map showing well locations and line of section.

The log expression for the N5.2 sand exhibits blocky/cylindrical motif (**Figure 4**). It has a sharp base with a gradational top, about 70.08 m thick, excellent reservoir quality, good lateral continuity and displaying a layer-cake reservoir architecture deposited probably as barrier bar or a channel sitting on a beach-barrier system (**Figure 4**). There are particles of glauconite in the ditch cutting sample indicating that the sediments were deposited within the marine environment.

3.2 Architecture and reservoir characteristics of the N5.2 sand

3.2.1 Architecture of the N5.2 reservoir

The root-mean-square amplitude, a seismic attribute generated over the seismic volume at different timeslices is more diagnostic in facies identification. Generally, the RMS amplitude is an expression of the square root of the average of the squares of the amplitude within certain window of the analysis, and it is related to the energy within the seismic trace. The root-mean-square (RMS) are useful in differentiating between lithology types. For instance, values of the trace with high amplitudes may indicate a highly porous lithology such as porous sand, which are potential high quality hydrocarbon reservoirs. It can also serve as a direct hydrocarbon indicator. Three different color bands are used to indicate amplitude values: yellow indicates the highest amplitude (100 RMS value); light brown to red (50–68 RMS values), and the light green to blue colors represent the least to medium amplitude points (0–18, and 20–30 RMS) (**Figure 5**). The N5.2 sand falls within the Beach-Barrier-shoreface (BBS) RMS trend that aligned parallel to paleo-coastline direction (**Figure 5**) indicating a shallow marine depositional architecture. Depth structure map also indicated similar trend showing a four-way dip closure having two culminations and depositional axis parallel to the major structure-building fault (**Figure 6**).

3.2.2 Reservoir characteristics of the N5.2 sand

Well log attributes show that the N5.2 reservoir is a massive, clean and thick (about 70.08 m thick) barrier/shoreface shallow marine sandstone with excellent reservoir quality. Pixel-based facies modeling using sequential indicator simulation

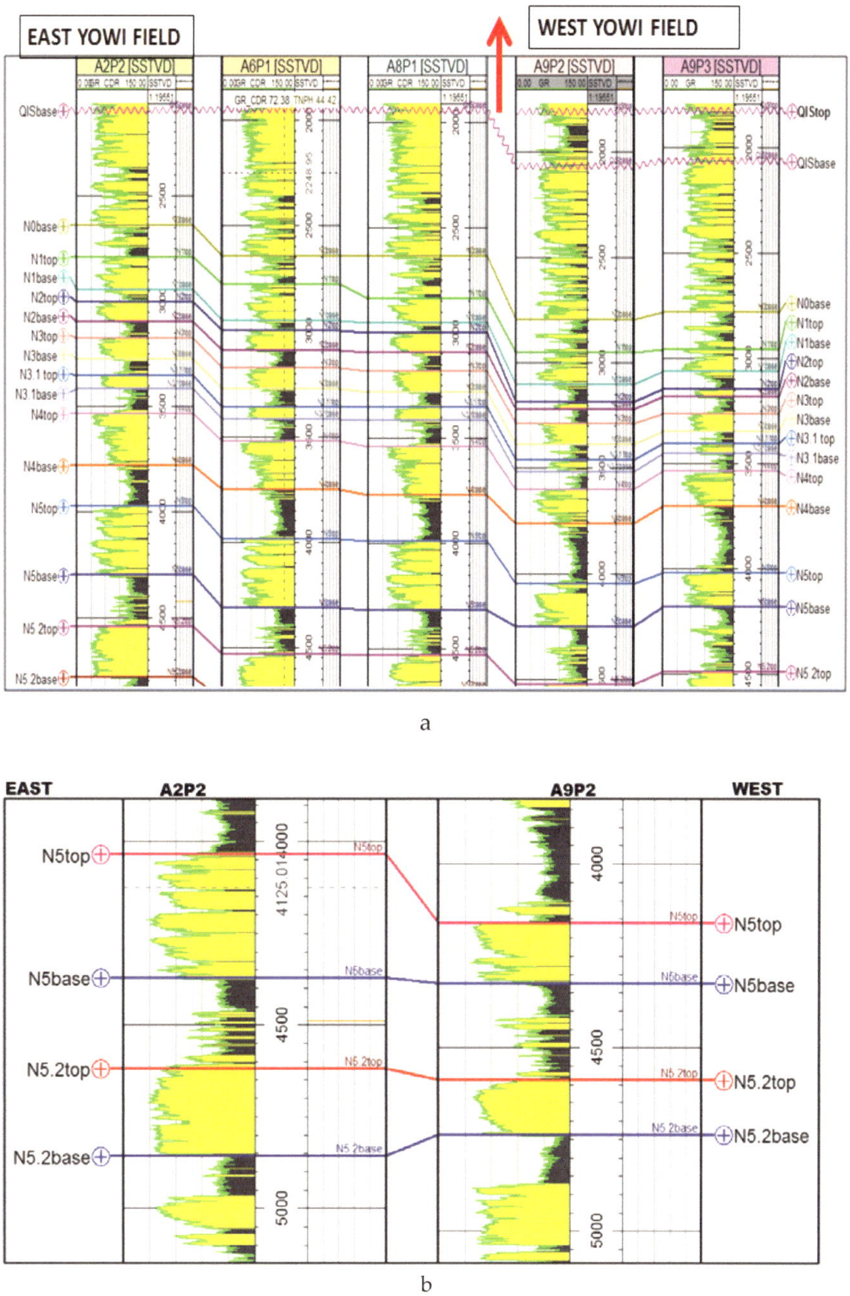

Figure 3.
(a) Well correlation of key reservoirs in the field and (b) well correlation of N5.2 reservoir across the field.

Reservoir	Sample description
(N5.2 reservoir) Well A2P2	Sandstone: medium to dark brown, oil saturated, clear, translucent, friable to loose quartz grains, very fine-fine grains, and predominantly very fine, sub-rounded to rounded, very well sorted, excellent porosity and permeability, contains mica with traces of glauconite, traces of carbonaceous speckles. Oil shows: medium brown oil stain, intense bright yellow fluorescence, instant blooming milky white cut fluorescence, light brown residue. Very strong hydrocarbon odor

Table 1.
Ditch cutting description of the N5.2 reservoir [15].

(SIS) algorithm for stochastic distribution of properties shows three distinct lithofacies (sand, siltstone and shale) in the reservoir across the field (**Figure** 7). These facies were defined based on log signatures, volume of shale cut-off, and

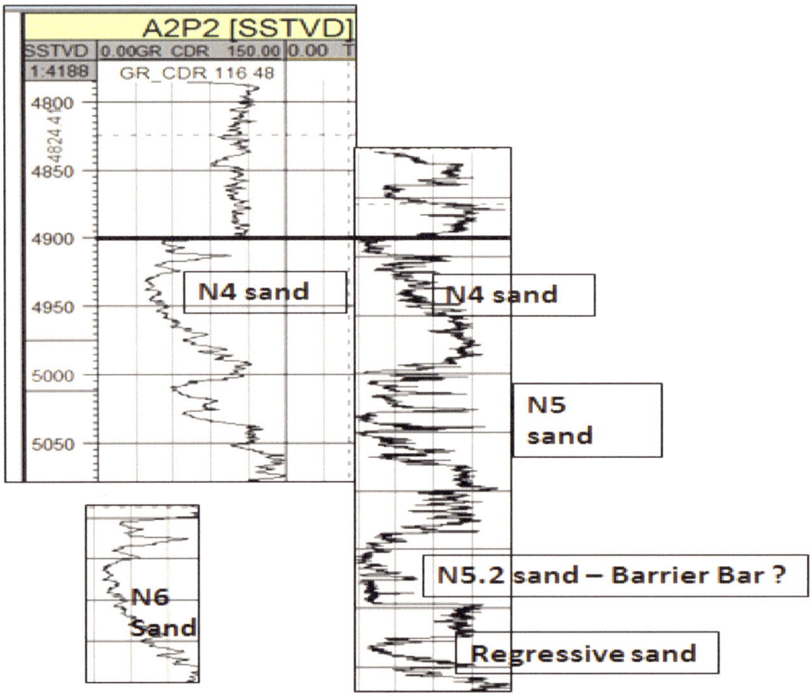

Figure 4.
Gamma ray log motifs showing inferred depositional environments of the reservoirs.

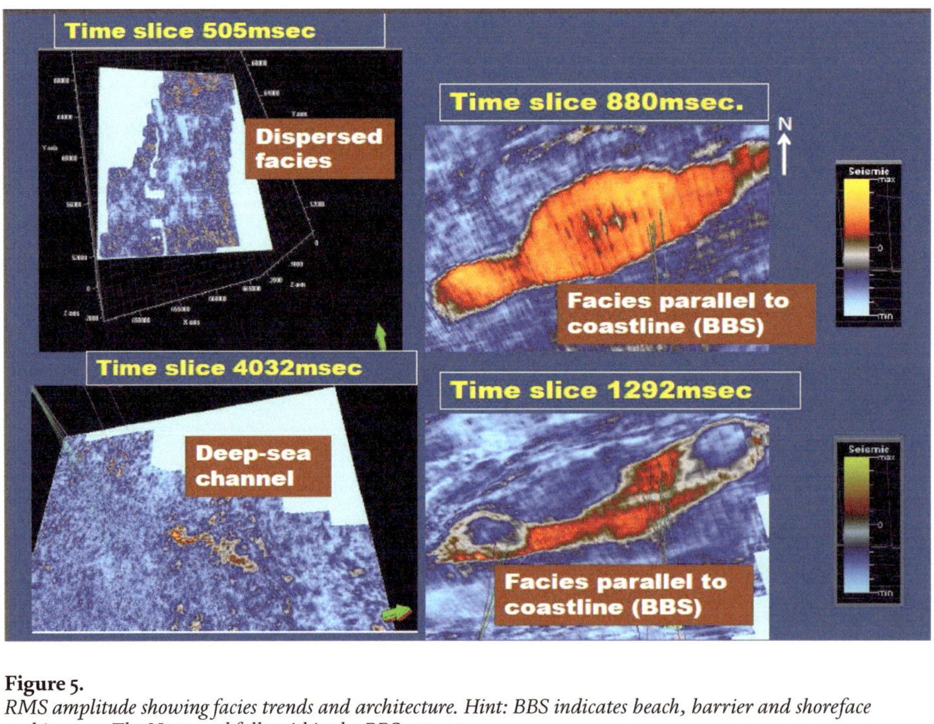

Figure 5.
RMS amplitude showing facies trends and architecture. Hint: BBS indicates beach, barrier and shoreface architecture. The N5.2 sand falls within the BBS category.

net-to-gross (NTG) calculations. The volume of shale (VSH), net-to-gross ratio (NTG) and total porosity (PHIT) from petrophysical evaluation were populated into the 3D grid facies model (**Figures 7** and **8**). Sand is the dominant lithofacies in

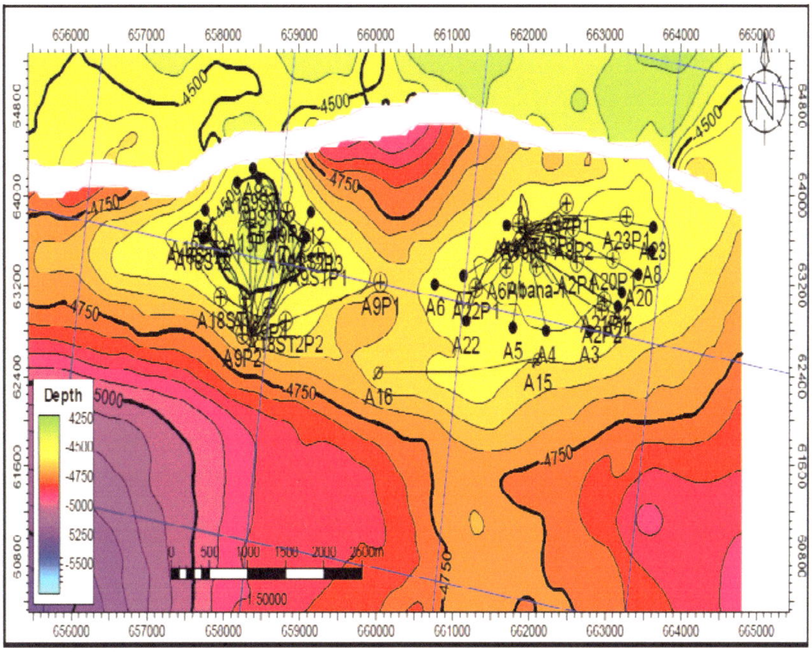

Figure 6.
Depth structure map for N5.2 reservoir in two structural culminations and a saddle.

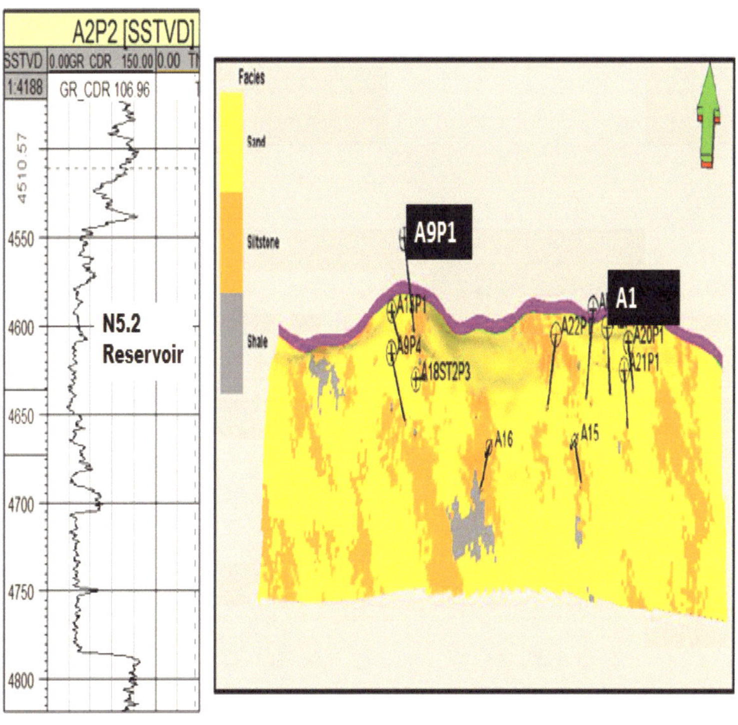

Figure 7.
Facies model showing lithofacies characteristics of N5.2 sand [25].

the reservoir (**Figure 5**) with over 83% NTG and negligible volume of shale (VSH) (**Figure 7**). This implies that the reservoir flow mechanism will be influenced by the properties of the sand facies.

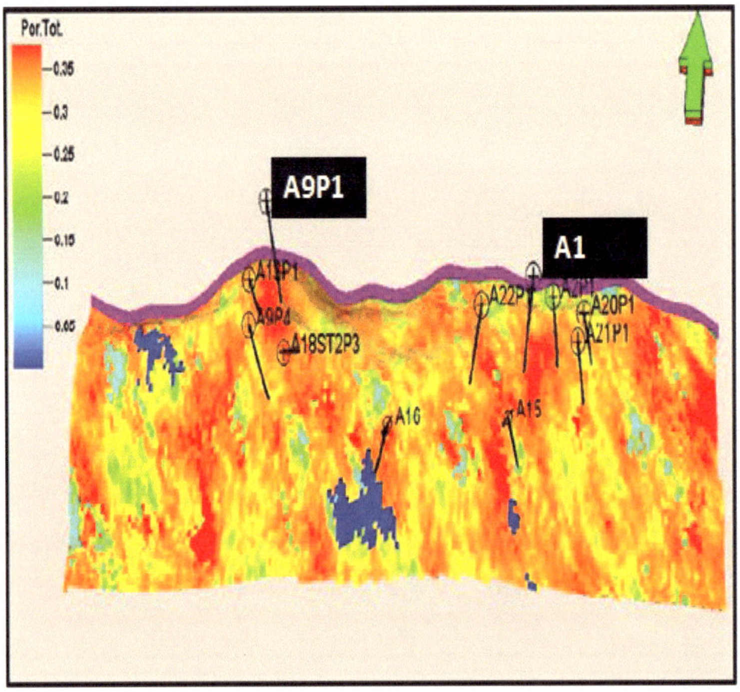

Figure 8.
Porosity model for N5.2 sand.

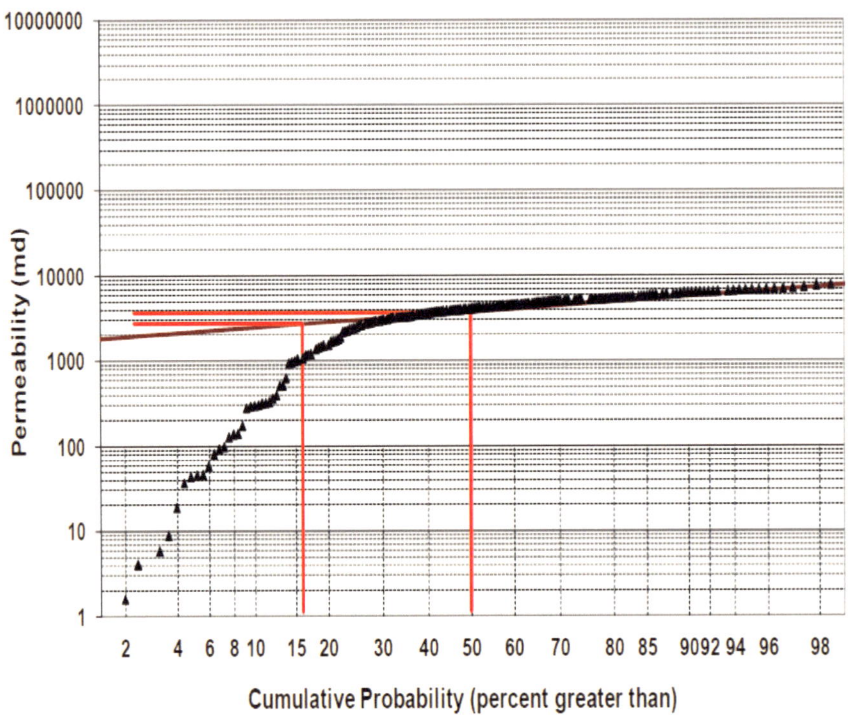

Figure 9.
Dykstra-Parson plot for heterogeneity test (N5.2 VDP = 0.3).

DOI: http://dx.doi.org/10.5772/intechopen.85517

3.2.3 Test for heterogeneity and flow unit characterization of the N5.2 sand

Dykstra-Parson's coefficient is an expression that measures the degree of variation and heterogeneity of a reservoir [26]. The variation in the values of the core permeability reflects the degree of heterogeneity in the reservoir. Rock samples with zero permeability values (shales) were not used since it is a logarithmic plot. The Dykstra-Parson plot performed for the core samples from the well section is shown in **Figure 9**. The Dykstra-Parson's number for the core samples is 0.30—which indicates a homogenous reservoir. Higher values of Dykstra-Parson indicate more heterogeneity of which one is the maximum number.

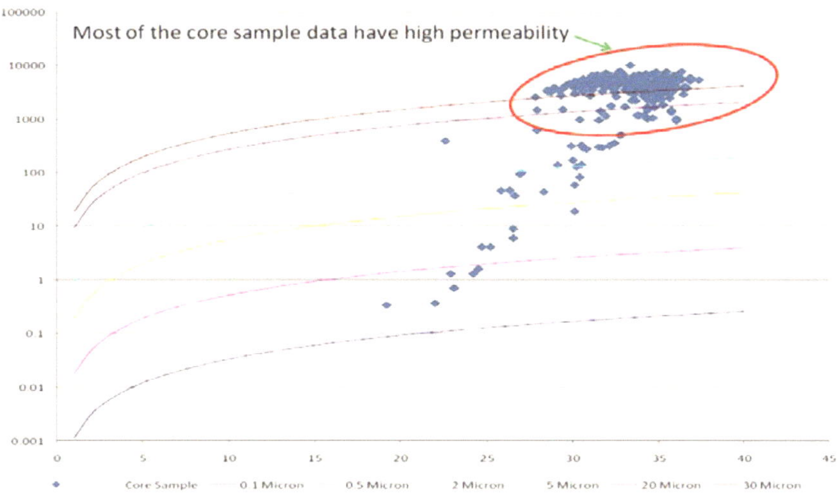

Figure 10.
Winland plot for the N5.2 sand.

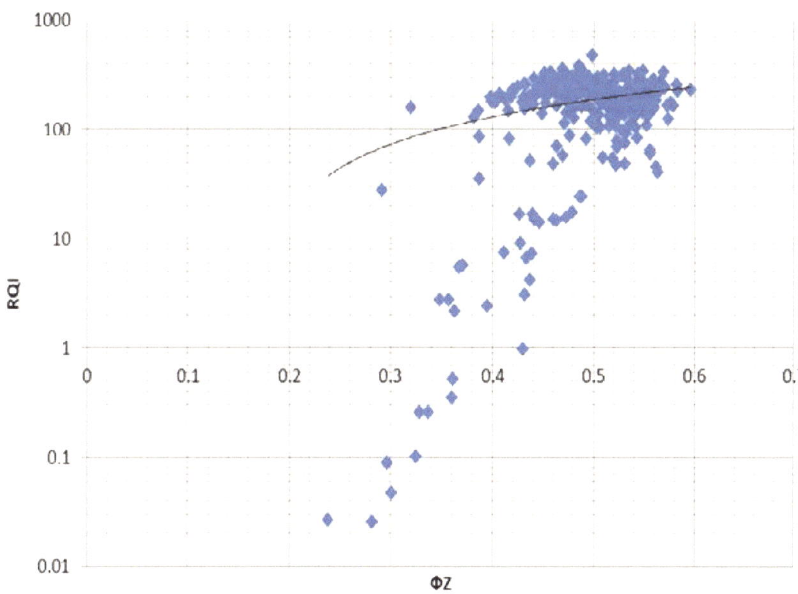

Figure 11.
RQI versus normalized porosity crossplot. Note that most of the samples have high quality index with RQI above 100. This is indicative of a high quality formation with high hydraulic potential.

The hydraulic capacity of the sand-body was well understood from plots generated from Winland and Rock Quality Index for flow zone characterization (**Figures 10** and **11**). Most of the samples plot within 1000 mD on Winland plot, showing a mega porous reservoir with a high quality index (**Figure 10**). Beside the presence of fractures that have created anisotropic condition in the homogenous geobody, excellent reservoir quality will enhance fluid flow in the reservoir [25]. Reservoir properties have major influence on reservoir fluids and the hydraulic behavior of the rock. It is important that these uncertainties are well understood because they are relevant to reservoir management decisions.

4. Possible causes of rapid production decline

4.1 Fracture network

Fractures are lineaments that occur in rocks which represent minor breaks in the natural order of the properties of the rock [27, 28]. They are evidences of the brittle failure of the rock due to lithostatic stresses initiated by tectonism and other geodynamic processes [29]. Network of fractures and faults have been identified in the area (**Figures 12–14**). They appear as short, disconnected and network of dark patches around the well area on seismic time slices [25, 28].

Several factors may be responsible for the development of these fractures. These could be due to changes in lithostatic pressure, geothermal stresses, hydraulic pressure and high drilling density in the field [25]. There is presently no infill drilling opportunity on the structure due to high drilling activity. The regional stress somewhat affects the orientation of the fractures as some are parallel to the axis of the growth faults, whereas some have multiple orientations especially around the well area (**Figures 12** and **13**). Non-uniform fracture distribution and heterogeneity in natural fractured reservoirs (NFR) make the development of water-cut asymmetrical and estimation of critical rate and breakthrough time will require fracture pattern modeling for proper understanding of fracture development around the producing wells. Consequently, tectonics, geothermal processes and human activity (drilling, well stimulation) could contribute to fracture generation. Fracture patterns and high vertical permeabilities created are also two important flow parameters that will allow for rapid non-uniform flow of water into the well. The N5.2 sand

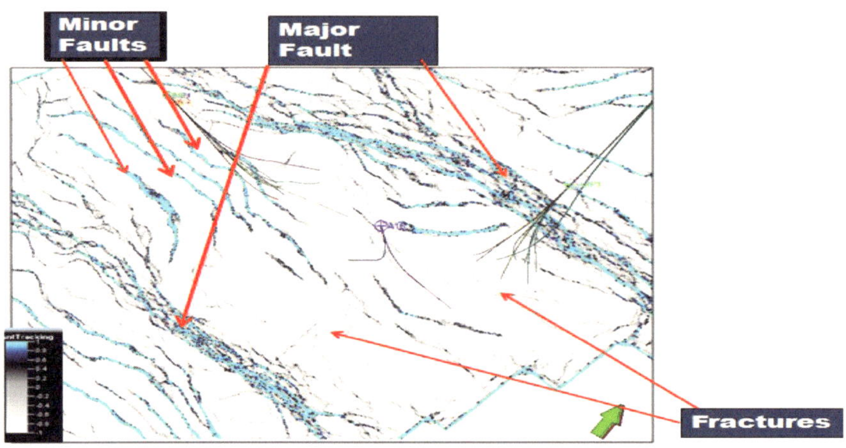

Figure 12.
Ant tracking at time-slice 1346.748 ms showing structural discontinuities.

falls between time slice 1346–1479 ms on the seismic section, where these fractures have been mapped, indicating that the reservoir is affected by the fractures. Five (5) CT-scanned core plugs taken from this interval also revealed the presence of massive vertical fractures (**Figure 14**) [25]. The vertical fractures radiate from the center of the core plugs to the edge, running around the internal circumference of the core as concentric rings (**Figure 14**).

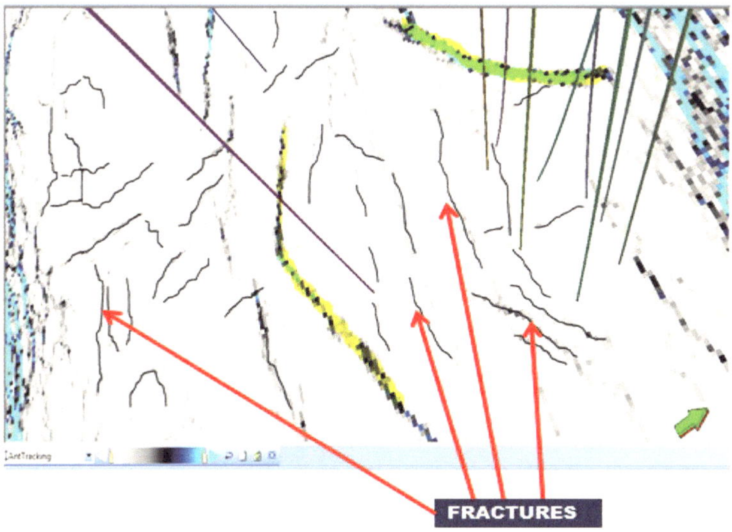

Figure 13.
Fractures highlighted to show orientations. Some of the fractures are aligned in the direction of the regional stress.

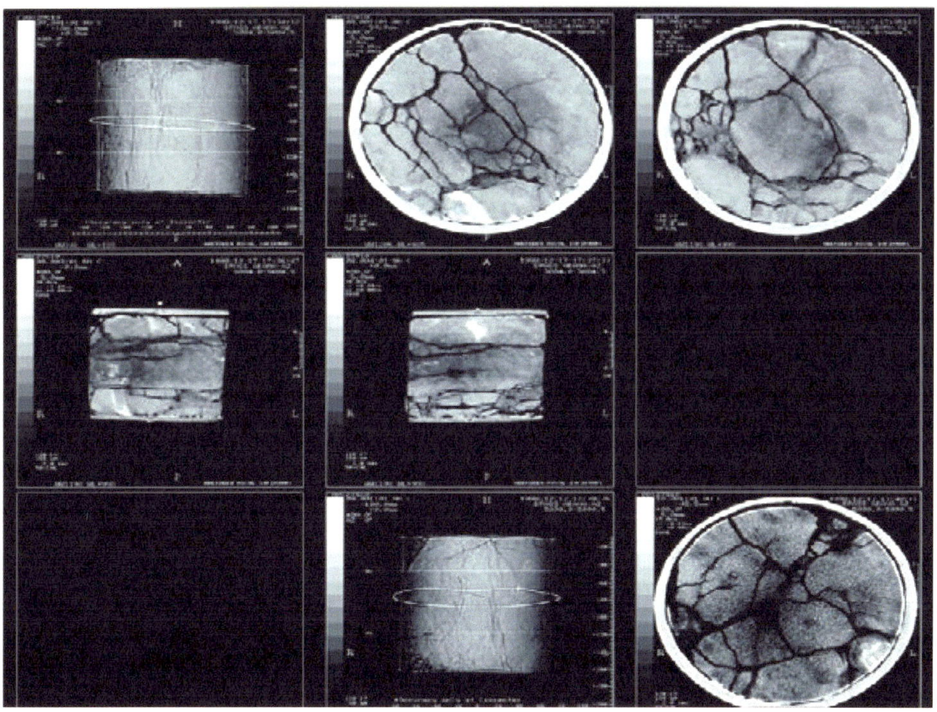

Figure 14.
Vertical fractures revealed in CT-scanned core sample p.

4.2 Production response

4.2.1 Effect of reservoir management on oil rate and basic sediment and water (BSW)

Production parameters considered include; Flowing Tubing Head Pressure (FTHP), Gas-Oil-Ratio (GOR), basic sediment and water (BSW) and choke size. The management of the choke has relative impact on oil rate and basic sediments and water (BSW). Practically, water production within the first 2 years of production was zero and peak production (over 4750 BOPD) was observed between the first 2 years (**Figures 15–17**).

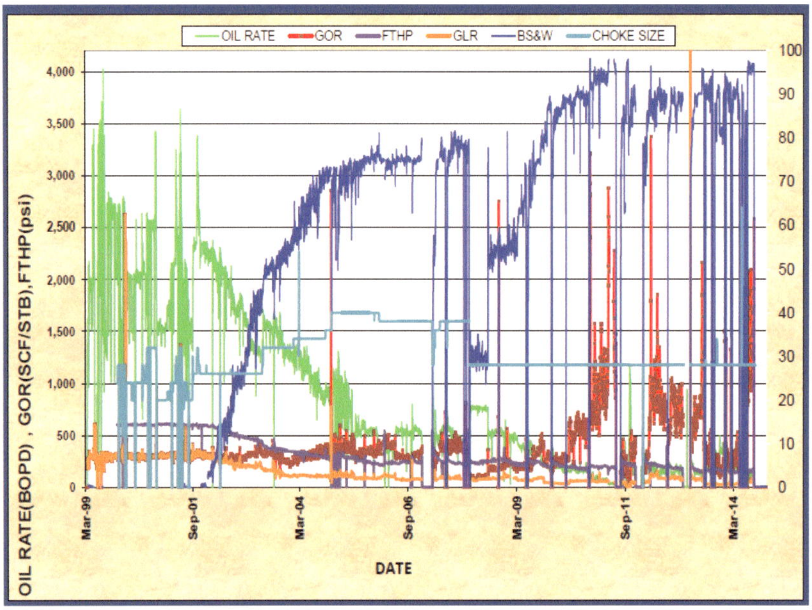

Figure 15.
The effect of choke management on oil rate and basic sediment and water for well A4.

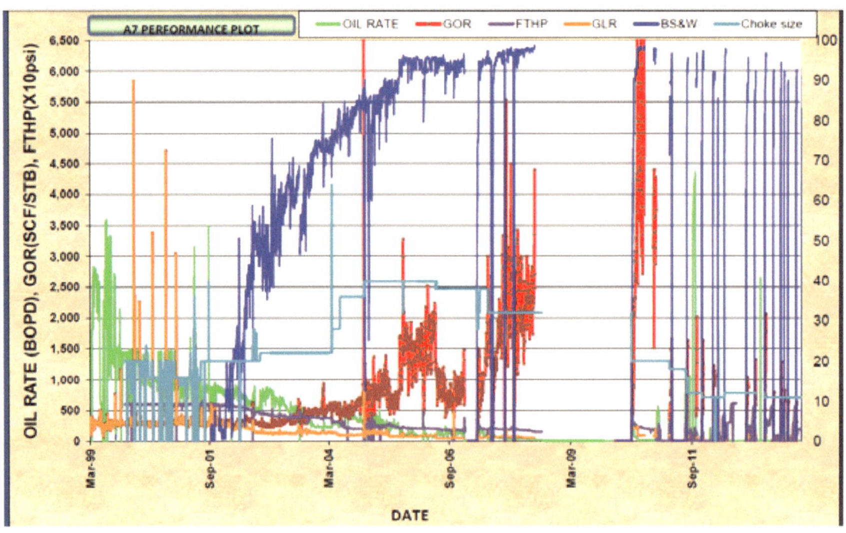

Figure 16.
The effect of choke management on oil rate and basic sediment and water for well A7.

There was high FTHP corresponding to periods of high oil rate (**Figures 15** and **18**) and choke size not exceeding 30 inches. Between September 2001 and March 2004, there was steady increase in choke from 33 to 60 (see **Figure 17**), with similar trend in the BSW. It is evident that the management of the reservoir also played a significant role on rapid fluid flow especially water.

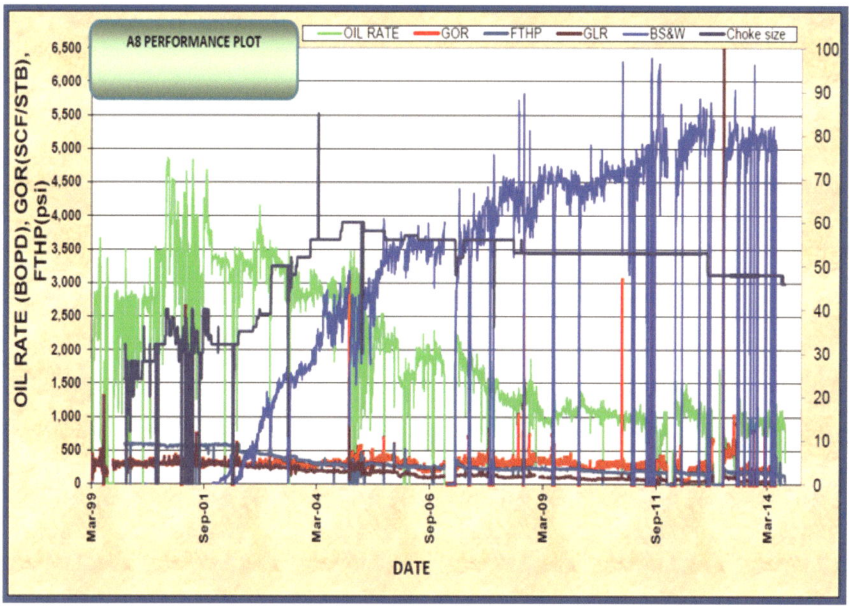

Figure 17.
The effect of choke management on oil rate and basic sediment and water for well A8.

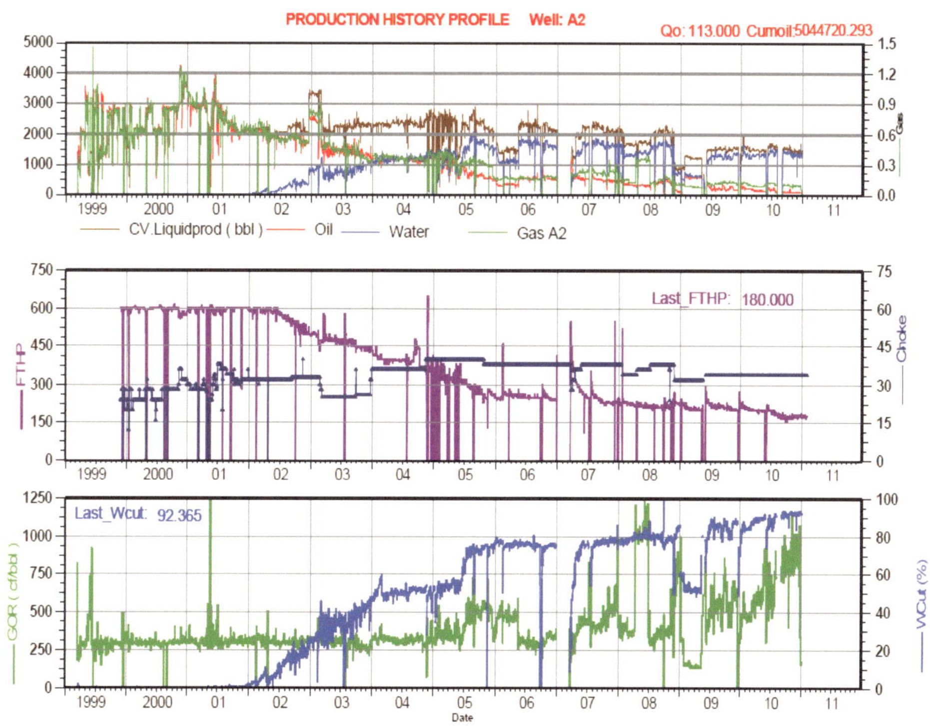

Figure 18.
Production history profile for well A2.

The continued water injection through the injector wells in the field could be another source of water production which has adverse effect on oil recovery especially for production wells proximal to the injectors.

The oil rate, Qo, as well as the liquid rate declined at the onset of water production in 2001 and afterwards liquid rate had remained constant at different choke sizes. The oil rate declined further as water cut increased (**Figures 15–18**). The total cumulative oil production as at January 2011 is about 67.18 MMSTB; this represents 43.91% recovery factor. However, there is still recoverable oil remaining in the reservoir, which can only be recovered if certain field operational conditions are considered.

5. Conclusions

The conclusions drawn based on the results of this study are as stated below:

1. The key geological factors that aided fluid mobility especially vertical flow include: environments of deposition, rock properties, lateral continuity of reservoirs, limited and negligible shale barriers in the N5.2 sand, and network of fractures. Reservoir properties of N5.2 sand show that the facies is highly homogenous and exerts a greater impact on fluid flow across the reservoir. The results of anisotropy due to permeability within facies on oil production increases with vertical flow potential. Fracture networks created anisotropic condition in the reservoir resulting in dominant vertical flow created by vertical fracture permeability. The major drive mechanism in the field is water; vertical fracture permeability would have created a high mobility ratio resulting in both the injected and displacing fluid breaking through earlier at the producer sand.

2. A significant coning effect is observed where there is rapid increase in choke size. It increases breakthrough time and decreases oil rate.

3. Increase in vertical fracture permeability resulted to early breakthrough and reduces oil rate after breakthrough.

4. Geological conditions (fractures and excellent reservoir quality) and poor reservoir management accounted for high fluid movement as well as high watercut in the field.

5. Seismic time-slicing methodology remains the approach to display the various seismic attributes, and useful in: identification of facies, selecting the best drilling locations, and measuring the architecture of a reservoir.

6. Water coning effect is a significant reservoir phenomenon that occurs in aquifer-supported reservoirs and is aggravated in a fractured reservoir system.

7. A synergy of a comprehensive geological and engineering investigation is a must at every stage in field development. It will provide full information needed to understand the subsurface.

Acknowledgements

The platform for this research was provided by the O.B. Lulu Briggs Chair in Petroleum Geoscience at the Institute of Petroleum Studies, University of Port

Harcourt, Nigeria, under the chairmanship of Prof. Minapuye I. Odigi, who diligently supervised the project. I appreciate the Late High Chief O.B. Lulu Briggs for his vision for research and capacity development at the University of Port Harcourt. My wife, Mrs. Perfect Suka and children, Ms. Victory and Virtue Suka gave me so much comfort and encouragement during the period of this study.

Conflict of interest

There is no conflict of interest in this research work.

Author details

Prince Suka Momta
Department of Petroleum Engineering and Development, Belemaoil Producing Limited, Nigeria

*Address all correspondence to: princemomta@yahoo.com

References

[1] Reijers TJA. Stratigraphy and sedimentology of the Niger Delta. Geologos. 2011;**17**(3):133-162

[2] Amajor LC, Agbaire DW. Depositional history of the reservoir sandstones, Akpor and Apara oilfields, eastern Niger Delta, Nigeria. Journal of Petroleum Geology. 1989;**12**(4):453-464

[3] Kamola DL, Van Wagoner JC. Stratigraphy and Facies Architecture of Parasequences with Examples from the Spring Canyon Member, Blackhawk Formation, Utah, Sequence Stratigraphy of Foreland Basin Deposits; Outcrop and Subsurface Examples rom the Cretaceous of North America. Tulsa, OK, United States: American Association of Petroleum Geologists; 1995. pp. 27-54

[4] Zhang J-J, Wu S-H, Fan T-E, Fan H-J, Jiang L, Cheng C, et al. Research on the architecture of submarine-fan lobes in the Niger Delta Basin, offshore West Africa. Journal of Palaeogeography. 2016;**5**. DOI: 10.1016/j.jop.2016.05.005

[5] Doust H, Omatsola E. Niger Delta. In: Edwards JD, Santogrossi PA, editors. Divergent/Passive Margin Basins. Tulsa: American Association of Petroleum Geologists Memoir. Vol. 48. 1990. p. 239-248

[6] Lisle RJ. Detection of zones of abnormal strains in structures using Gaussian curvature analysis. The American Association of Petroleum Geologists Bulletin. 1994;**78**:1811-1819

[7] Tyler N, Ambrose WA. Facies architecture and production characteristics of strandplain reservoirs in the Frio Formation, Texas. In: Report of Investigation No. 146. Austin Texas: Bureau of Economic geology, University of Texas at Austin; 1985. p. 78713

[8] Rassi C, Hentz TF. Production prediction and reservoir characterization of systems tracts of fourth-order sequence in the Miocene offshore Louisiana (abs). American Association of Petroleum Geologists Annual Convention, Offshore program. 2003;**10**:A164

[9] Fitch PJR, Jackson MD, Hampson GJ, John CM. Interaction of Stratigraphic and Sedimentological Heterogeneities with Flow in Carbonate Ramp Reservoirs: Impact of Fluid Properties and Production Strategy. Lyell collection. EAGE/The Geological Society of London; 2014. Available from: http://pg.lyellcollection.org

[10] Tveranger J, Braathen A, Skar T, Skauge A. Centre for integrated petroleum research-research activities with emphasis on fluid flow in fault zones. Norwegian Journal of Geology. Tidsskrift. 2005;**85**:63-71

[11] Adeogba AA, McHargue TR, Graham SA. Transient fan architecture and depositional controls from near-surface 3D data, Niger Delta continental slope. American Association of Petroleum Geologists Bulletin. 2005;**89**(5):627-638

[12] Arochukwu E. Assessing the Impact of Reservoir Architecture on Secondary Recovery: A Case Study of "Century" Field, Niger Delta. NAPE Monthly Technical Meeting, Port Harcourt. 2014

[13] Weber KJ. Sedimentological aspects of oil fields in the Niger Delta. Geologie en Mijnbouw. 1971;**50**:559-576

[14] Momta PS, Omoboh JO, Odigi MI. Sedimentology and depositional environment of D2 sand in part of greater Ughelli Depobelt, onshore Niger Delta Nigeria. American Journal of Engineering and Applied

Sciences. 2015;**8**(4). DOI: 10.3844/ajeassp.2015.556.566

[15] Momta PS, Odigi MI. Reconstruction of the depositional setting of tortonian sediments in the Yowi field, shallow offshore Niger Delta, using wireline logs. American Journal of Geosciences. 2016;**6**(1). DOI: 10.3844/ajgsp.2016.24.35

[16] Osinowo OO, Ayorinde JO, Nwankwo C, Ekeng OM, Taiwo OB. Reservoir description and characterization of Eni field offshore Niger Delta, southern Nigeria. Journal of Petroleum Exploration and Production Technologies. 2017. DOI: 10.1007/s13202-017-0402-7

[17] Oyedele KF, Ogagarue DO, Mohammed DU. Integration of 3D seismic and well log data In the optimal reservoir characterisation of EMI field. Offshore Niger Delta oil province, Nigeria. American Journal of Scientific and Industrial Research. 2013. DOI: 10.5251/ajsir.2013.4.1.11.21

[18] Adeoti L, Onyekachi N, Olatinsu O, Fatoba J, Bello M. Static reservoir modeling using well log and 3-D seismic data in a KN field, offshore Niger Delta, Nigeria. International Journal of Geosciences. 2014;**2014**(5):93-106. Available from: http://www.scirp.org/journal/ijg. DOI: 10.4236/ijg.2014.51011

[19] Prather BE. Controls on reservoir distribution, architecture and stratigraphic trapping in slope settings. Marine and Petroleum Geology. 2003;**20**:529-545

[20] Richards M, Bowman M. Submarine fans and related depositional systems II: Variability in reservoir architecture and wireline log character. Journal of Marine and Petroleum Geology. 1998;**15**:821-839

[21] Howell JA, Flint SS. Siliciclastics case study: The book cliffs. In: Coe A, editor. The Sedimentary Record of Sea-Level Change. Cambridge, UK: Cambridge University Press; 2003. pp. 135-208

[22] Kirschbaum MA, Hettinge RD. Facies Analysis and Sequence Stratigraphic Framework of Upper Campanian Strata (Neslen and Mount Garfield Formations, Bluecastle Tongue of the Castlegate Sandstone, and Mancos Shale), Eastern BookCliffs, Colorado and Utah. U.S. Geological Survey Digital Data Series DDS-69-G. Denver, Colorado: Central Region; 2004. ISBN: 0-607-90854-5

[23] Slatt RM. Stratigraphic Reservoir Characterization, Handbook of Petroleum Exploration and Production. The Netherlands: Elsevier; 2006. pp. 8-30

[24] Wright IP, Malcolm FF, George ZJ. The Abana Field, Nigeria, from Discovery to Recovery: A Case Study. In: American Association of Petroleum Geologists Search and Discovery Article #90923@1999 International Conference and Exhibition. Birmingham, England; 1999

[25] Momta PS, Odigi MI. Impact of structural discontinuities on fluid flow and production behaviour: Case study of the Yowi field, offshore Niger Delta, Nigeria. Society of Petroleum Engineers. 2015. DOI: 10.2118/178260

[26] Dykstra H, Parsons RL. The Prediction of Oil Recovery by Waterflooding in Secondary Recovery of Oil in the United States. 2nd ed. Washington, DC: American Petroleum Institute. 1950. pp. 160-174

[27] Djebbar T, Donaldson EC. Petrophysics-Theory and Practice of Measuring Reservoir Rock and Fluid Transport Properties. Burlington: Gulf Professional Publishing is an imprint of Elsevier; 2004. pp. 488-548

[28] Momta PS, Odigi MI. Drilling deeper to access more hydrocarbon potentials in the Yowi Oilfield, Offshore, Niger Delta, Nigeria. In: SPE Nigeria Annual International Conference and Exhibition; 4-6 August; Lagos, Nigeria. Society of Petroleum Engineers; 2015. DOI: 10.2118/178261

[29] Djebbar T, Donaldson EC. Theory and Practice of Measuring Reservoir Rock and Fluid Transport Properties. USA: Gulf Professional Publishing, Elsevier; 2012. p. 950

Chapter 3

"Geo-archives of a Coastal Lacustrine Eco-system": Lake Bafa (Mediterranean Sea)

Özlem Bulkan, Bilgehan Toksoy Ediş and M. Namık Çağatay

Abstract

A hypothetical novel, which aims to summarize the whole geological history of the coastal area surrounding the Mediterranean Sea, probably contains a sum of intensive and impressive topics, such as tsunamis, storms, earthquakes, volcanic activities, human-nature interactions, and their products. These abrupt geo-event changes (e.g., water chemistry fluctuations) remark a dynamic nature of this unique coastal area. Paleolimnological studies of a coastal lacustrine archive (i.e., Lake Bafa, Turkey), associated with syngenetic deposits accumulated in neighboring geological settings (i.e., swamp, deltaic, lagoon, marine), has allowed us to reconstruct the local geological history. Following this hypothesis, we aimed at investigating the paleoenvironmental establishment of the Lake Bafa and surrounding coastal area. Lithologic and geochemical investigations of the lacustrine ("BAF37:4.2 m) core and surrounding swamp ("BS":12 m) sediments supplied us an excellent geo-archive, continuously accumulated during the last 4.5 ky. Following conclusions are provided concerning the main depositional stages: Recent swamp-lacustrine separated stage (S-I: last 0.8 ky), lagoon stage (S-II.: 0.8–1.75 ky BP), marine-river interaction stage (S-III: 1.75–2.7 ky BP), and marine-dominated stage (S-IV: 2.7–4.5 ky BP). Our observations indicate that ecosystem characteristics of the basin have been mainly controlled by the hydroclimate and geotectonic processes.

Keywords: Lake Bafa, coastal lake, Holocene environment, sediment geochemistry, isotope geochemistry, Mediterranean Sea

1. Introduction

The marine and terrestrial geological settings influence each other in the coastal areas [1, 2]. The Aegean Sea and its onshore areas have raised a substantial interest of the geologists and geomorphologists, since this region is under a north–south tectonic extension, with the formation of hosts and grabens and occurrence of high seismic activity [3]. These very active tectonics, together with postglacial sea level rise, modified the geomorphological evolution of the Aegean coastal area, driving intensive sediment transport along the main river systems (e.g., Büyük Menderes), high amount of sediment accumulation with delta formations, and progradational deltaic processes [4, 5]. Accordingly, the late Quaternary geomorphological development of the eastern coast of the sea witnessed the formation of various inland or transitional basins.

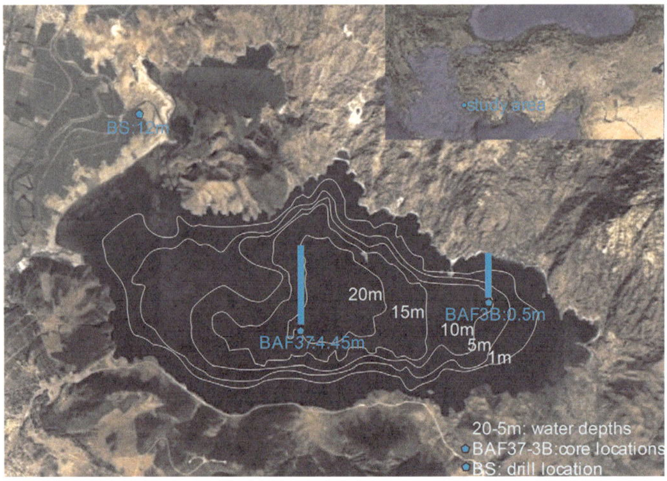

Figure 1.
Lake Bafa bathymetric map showing the piston core and drill core locations.

As a Middle to Late Holocene age coastal lacustrine basin formed in this specific geological area, Lake Bafa shows a complicated paleoenvironmental and paleoecological history [6, 7] (**Figure 1**). The lake evolved by the progradation of the Büyük Menderes River (Maiandros, Maeander) delta and the closure of the former Latmian Gulf during the late glacial-Holocene transgression [6]. It was subjected to ecological changes from a coastal marine inlet to a lagoon and finally to a completely isolated lake [6, 8, 9]. From this point of view, Lake Bafa provides an important sedimentary archive of these ecosystem changes and associated physical–chemical shifts, which are the main objectives of this study. To achieve these objectives, we carried out lithological descriptions and sedimentological and geochemical analyses of various sediment cores from the lake. The use of specific element concentrations and their ratios and organic geochemical analyses allowed us to reconstruct the past organic matter productivity-preservation rates, the water column chemistry, and the clastic material supply signals [10–12].

2. Samples and methods

An 11.9-m-long borehole (BS) was drilled on a swamp area, near the northwestern boundary of Lake Bafa, during a field survey in December 2012 (**Figure 1**). Subsequently, cores BAF-3B (0.5 m) and BAF37 (4.1 m) were retrieved from the central and eastern parts of the lake, using hammer and Kajak coring methods, respectively. The borehole section and sediment cores were lithologically described, systematically subsampled, and dried by using the freeze-drying method. Prepared mixtures of the selected 34 subsamples were analyzed as $LiBO_2/Li_2B_4O_7$ flux by ICP-MS in ACME laboratories, Canada. AMS 14C age determinations of selected *Cerastoderma glaucum* shells (**Table 1**) were analyzed in Beta Analytical Laboratories [8, 9, 13]. Calibration of the samples was calculated in Beta Analytical Laboratories, using one of the databases associated with the 2013 INTCAL program, using the reservoir age correction of 400 years [14, 15]. However, we realize that the reservoir age for samples representing the isolated stage of Lake Bafa (e.g., sample BAF 37/2, 72–73) may be different and that this deserves further investigation. Therefore, we submitted here the carbon dating data either in measured values and calibrated data (**Table 1**). Total organic carbon (TOC) analysis and Rock Eval

Submitter Nr.	Material pretreatment	Measured age	13C/12C	Conventional age	2 Sigma calibration
BAF37/2 72–73cm	(Shell): acid etch	1220 +/− 30 BP	−3.9 o/oo	1570 +/− 30 BP (1457 ± 70 adjusted for local reservoir correction)	Cal AD 780 to 1065 (Cal BP 170 to 885)
BAF37/P3–96	(Shell): acid etch	2250+/− 30 BP	−1.6	2345 BP+/− 30 BP	Cal BC 360 (Cal BP 2310)
BAF37/4 19–20cm	(Shell): acid etch	1980 +/− 30 BP	−3.6 o/oo	2330 +/− 30 BP (2217 ± 70 adjusted for local reservoir correction)	Cal BC 30 to AD 295 (Cal BP 1980 to 1655)
BS-FM/9 K-BS 14–7 cm	(Shell): acid etch	2710	−3.1	3070 +/− 30 BP (2957 ± 70 adjusted for local reservoir correction)	Cal BC 910 to 655 (Cal BP 2860 to 2605)
BS-9 K- 65–67 cm	(Shell): acid etch	2450+/− 30 BP	−4.7	2780 +/− 30 BP (2667 ± 70 adjusted for local reservoir correction)	Cal BC 645 to 230 (Cal BP 2595 to 2180)

Table 1.
AMS radiocarbon ages, calibrated ages, and description of the related sediment samples.

Pyrolysis VI measurements are performed in Turkish Petroleum Cooperation Laboratories.

3. Geological setting and limnology

From a geologic point of view, Western Anatolia and the Aegean Regions represent a broad extensional zone [16], stretching from Bulgaria to the north to the Hellenic arc to the south [17]. The Western Anatolian region is characterized by several approximately E-W trending, subparallel, normal fault zones, bordering a set of grabens and intervening horst blocks. The Lake Bafa is located on the Büyük Menderes Graben zone [18], which is a seismically active depositional basin [19]. The graben was opened in the Paleozoic–Mesozoic rocks of the Menderes Massif and Lycian Nappes during the Early Miocene [19–21]. The catchment mainly consists of metamorphic bedrocks belonging to the Menderes Massif. The basin-fill deposits, partly Early Miocene lacustrine limestones and the overlying units of Pliocene and Quaternary clastics, overlie the bedrocks [19]. The Lake Bafa formed as an alluvial set lake in the Western Anatolia because of the closure of the ancient Latmos Gulf, caused by the delta progradation of the Büyük Menderes River [6, 16]. The Holocene deposits of the Büyük Menderes River form an alluvial delta plain, separating the Lake Bafa from the Aegean Sea at around A.D. 1500 [6, 22]. The present- day lacustrine basin is currently at 2 m above sea level (masl) (**Figure 1**). The modern basin of the lake has a surface area of 315 km^2, a volume of 692 hm^3, and a maximum depth of 20 m. The lake is oligo-mesotrophic with annual average values of total nitrogen 0.45 mg/L, total phosphorus 1.3 mg/L, and total dissolved oxygen 7.49 mg/L [23, 24]. Additional to surface inflows, the main recharging inlet is Büyük Menderes River [22, 25].

4. Lithostratigraphy

Based on visual observations, such as lithology, color, water content, grain size distributions, and fossil content, sediment core and swamp section were subdivided into several lithostratigraphical units (BAF37, five litho-zones; BS, four litho-zones).

4.1 Lithology of the sediment–water interface core BAF-3B

The lithology of the sediment–water interface core "BAF-3B" from the north-eastern part of the lake predominantly contains fine-grained sediments, with averages of 18% sand, 34% silt, and 40% clay. The uppermost 0.15 m of the core contains gray and greenish-gray homogeneous sandy silty clays. Sediments from the lowermost parts were characterized by light to dark gray laminated sandy silty clay layers, enriched with small bivalve shell fragments [8].

4.2 Lithology of the core BAF37

The sedimentary section of core BAF37 consists of five main lithostratigraphic units (**Figure 2**). The uppermost 1.2 m of the sediment section is a greenish-olive gray homogeneous sandy, clayey silt. A 15 cm thick, black, banded, homogeneous clayey silt layer follows this unit downward. The third unit (UII:1.35–1.90 m) is a light bluish gray to yellow, laminated clayey silt. The following unit (UIV:1.90–3.56 m) consists of intercalations of bluish gray homogeneous sandy silty clay and olive-green clayey silt and clay. The lowermost unit below 3.56 m (UV) has variable lithological layers changing from homogeneous to laminated and from sands to clays, with beige and light gray to black colors. In detailed lithological observations including the grain size composition data (averages of 4% sand, 68% silt, and 28% clay), this unit is composed of predominantly fine-grained sediments [8, 9].

4.3 Lithology of the borehole (swamp) section (BS)

The swamp section is divided into four main lithological units. The uppermost 2.7 m of the sequence (Us I) is mainly characterized by brown sandy clay (**Figure 3**). Unit "Us I" includes a topsoil formation and is rich in plant residues in the uppermost part. Lowermost parts include a well-sorted sand layer and partly oxidized clay layers. The second unit (Us II; 2.7–7.2 m depth) contains laminated,

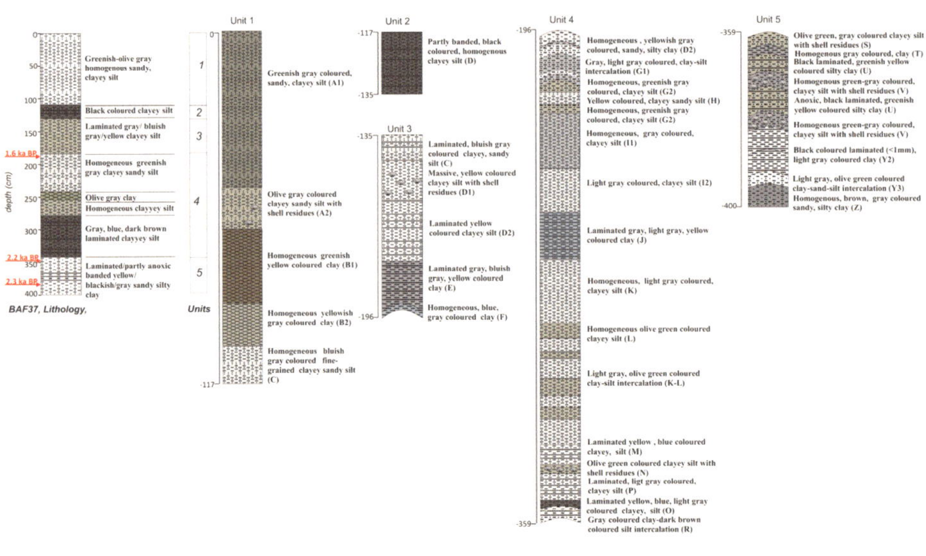

Figure 2.
Lithostratigraphic description of core BAF37.

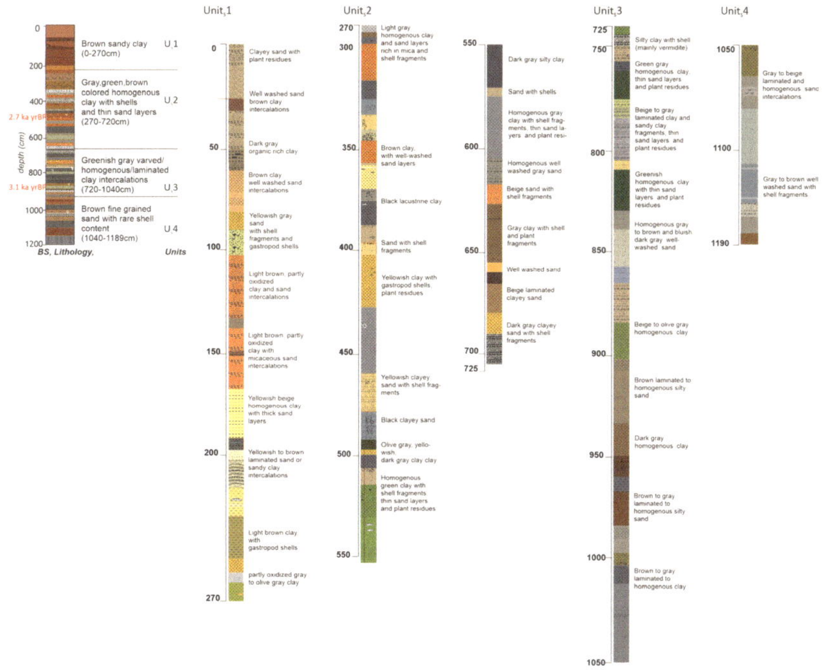

Figure 3.
Lithostratigraphic description of the swamp section (drill core BS).

partly organic-rich or oxidized clay layers, enriched in gastropod shells and shell fragments. The third unit (Us III; 7.2–8.7 m) is characterized mainly by a gray to brown, homogeneous clay, which is partly interrupted by either thin laminated clay layers or sandy, silty clay layers. It is enriched in Vermetidae or bivalve shell fragments. The lower parts of this unit (8.9–10.4 m) consist predominantly of homogeneous sand layers, containing abundant bivalve shells. The lowermost unit (Us IV; 10.4–11.9 m) is a gray, well-sorted homogeneous sand, with the abundant rock fragments and bivalve shells.

5. Chrono-stratigraphy

A total of five radiocarbon ages measured on mollusk shells were considered for the chronology (**Table 1**) [8, 9]. Three *Cerastoderma glaucum sp. shells* from three different depths in core BAF-37 were radiocarbon dated. A single valve of *Cerastoderma glaucum sp. from* 1.98 m depth provided a conventional age of 1570 (+/− 30) BP. Additional two AMS radiocarbon dates from single valves of *Cerastoderma glaucum sp. from* 3.38 m and 3.83 m core depths yielded 2250 (+/−30) and 2330 (+/−30) year BP. The 14C ages show a regular increase with depth. Furthermore, these age estimates indicate a low sedimentation rate during the deposition of lake sediments (0.24 cm/year); this rate is in agreement with the results of previous measurements [22]. Two AMS radiocarbon dates from single *Cerastoderma glaucum sp. shells* collected at the 4.81 m and 7.53 m depths of swamp section (BS) yielded conventional dates of 2780 (+/− 30) and 3070 (+/− 30) BP, respectively. According to the radiocarbon dates, together with visual observations, the swamp and lake sequences represent continuous sedimentary records of the last 4.5 and 2.5 cal. ka years, respectively.

6. Chemo-stratigraphy

Chemical characterization of the sediments (BAF-3B; BAF37; BS) investigated applying ICP-Ms analysis and revealed abundances of selected elements (Al, K, Ti, Zr, Rb, Fe, Mn, Ni, V, Cu, Pb, Zn, Mg, Ca, Na, P, Ba, Sr). A selected statistical method is also applied using geochemical data, namely, "factor analysis" (FA). Accordingly, eigenvalues of I and II factors were determined from 16 variables (**Table 2**). Congruent factor load values indicate either the same geological sources or element enrichment processes [26]. However, several element contents (Cu, V, Pb, Zn) were below the detection limits (*bdl*) in the sediment–water interface sediments (core: BAF-3B). These values were not applied for FA approaches.

6.1 Chemical composition of core BAF37

Unit I has relatively high Ca, Ni, Fe, and Zr concentrations; low contents of Si, K, Ti, Mn, and Rb; and low ratios of Mn/Fe, V/V + Ni, and Mg/Ca (**Figure 4**). Contrarily, a tendency through the increased Mn/Fe and Mg/Ca ratios suggested for the sediments belonging to units II to IV. Overall downcore characterization also reflects increasing values of Si, K, Ti, Fe, Ni, Zn, Ca, and Zr in units III and IV. These elements correlated oppositely with declining Ca values, indicating the carbonate dilution effect. Especially, the lowermost unit (Unit V) is characterized by the obvious fluctuations of the Ca, Fe, Si, K, Ti, Rb, and Fe. In addition to element concentrations, cross-correlations of V/Cr–U/Th, Ni/Co–U/Th, V/V + Ni–U/Th,

Cores	BAF 3B		BAF37		BS	
	factor 1	*factor 2*	*factor 1*	*factor 2*	*factor 1*	*factor 2*
Al	**−0.9**	−0.1	**−0.7**	0.6	**−0.7**	−0.6
Fe	**−0.9**	0.0	**−0.9**	0.3	**−1.0**	−0.2
Mg	**−1.0**	0.0	**−0.9**	0.5	**−0.8**	0.2
Ca	0.3	**0.9**	−0.2	**0.8**	**−0.7**	0.6
Na	0.7	−0.7	0.4	0.6	**0.9**	−0.4
K	**−0.9**	−0.2	**−0.9**	0.3	−0.7	−0.7
Ti	**−0.9**	−0.3	**−0.8**	0.6	**0.8**	−0.5
P	**0.7**	−0.2	0.0	**0.9**	**0.9**	−0.1
Mn	**−0.8**	0.5	−0.7	−0.2	**−0.9**	0.2
Ni	**−0.9**	−0.1	**−0.8**	−0.4	**−0.9**	0.2
Ba	**−0.9**	0.3	**−0.8**	−0.3	−0.5	**−0.7**
Sr	**0.8**	0.3	0.7	0.4	−0.5	0.6
V	*bdl*	*bdl*	**−0.8**	−0.4	**−0.9**	−0.3
Cu	*bdl*	*bdl*	−0.5	−0.5	**−0.9**	−0.1
Pb	*bdl*	*bdl*	0.4	−0.3	**−0.9**	0.0
Zn	*bdl*	*bdl*	−0.6	−0.6	**−0.9**	−0.3
Expl.Var	8.4	1.8	7.5	4.3	10.7	2.9
Prp.Totl	0.7	0.2	0.5	0.3	0.7	0.2

Table 2.
Factor loads of 16 selected elements (bold numbers indicate high positive and high negative factor loadings).

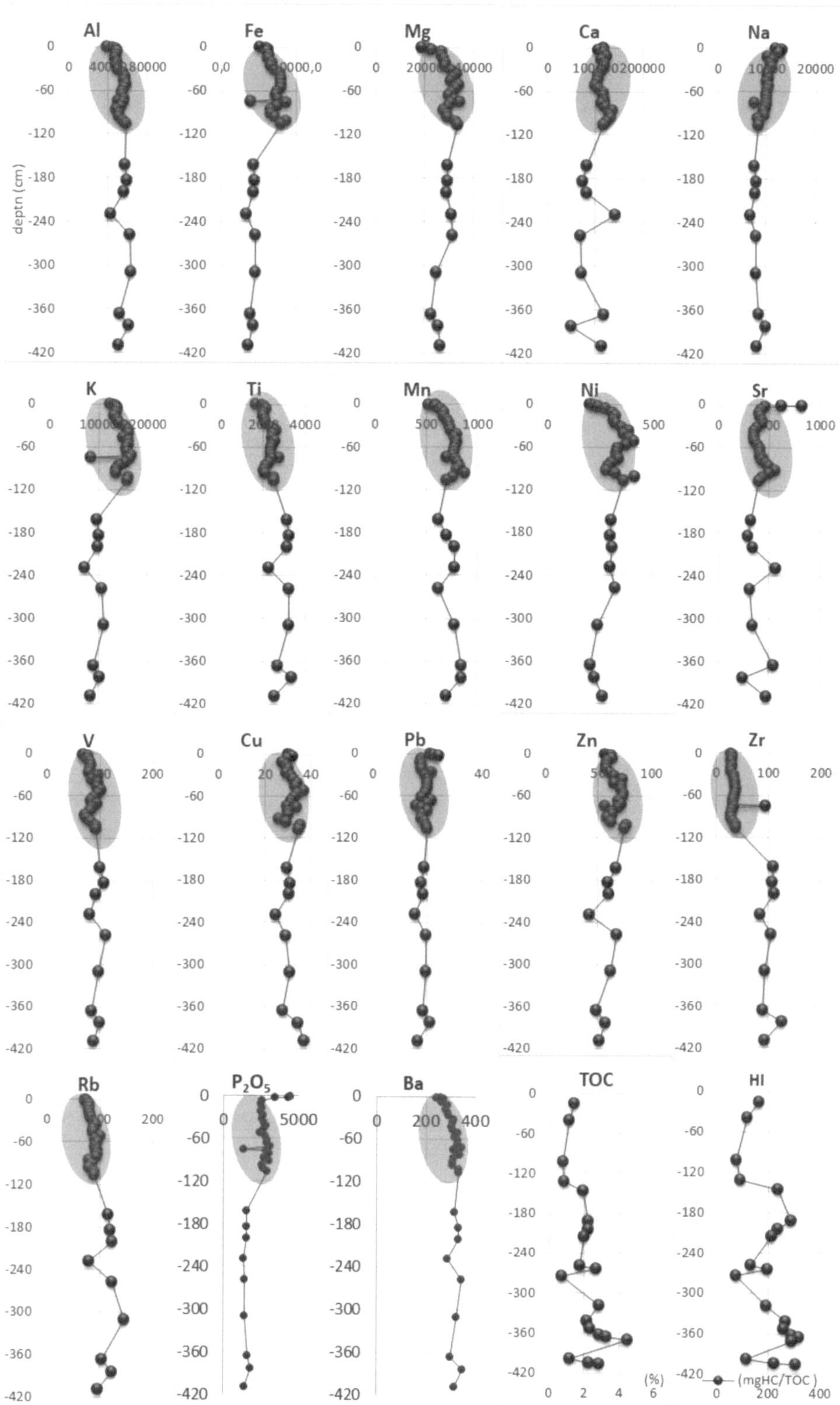

Figure 4.
Concentrations of selected element and element oxides (in ppm) in BAF37 core sediments (in ppm), TOC% contents, and HI values in BAF37 sediments (blue-colored area represents values measured for sediment–water interface phase and long core sediments, jointly).

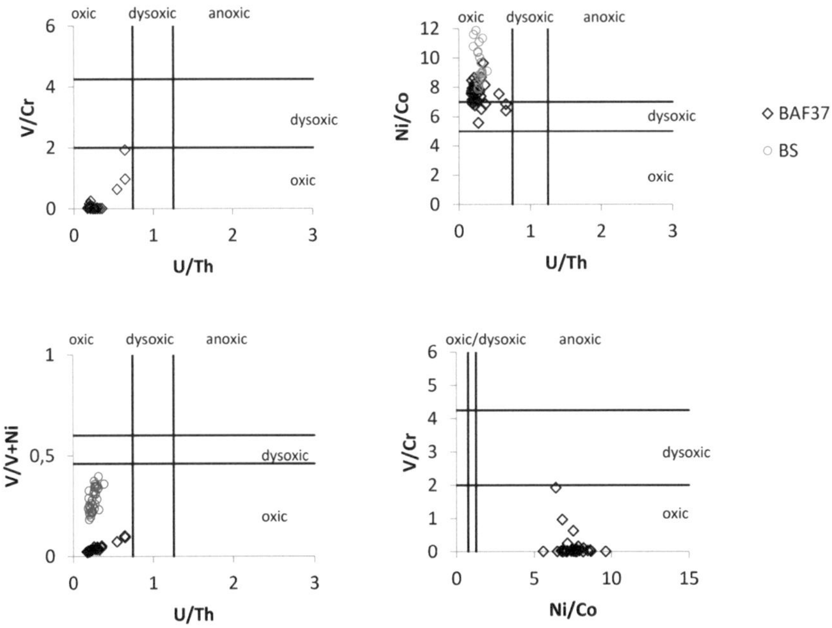

Figure 5.
Redox conditions of the lake bottom waters and the adjacent swamp environment.

and V/Cr–Ni/Co were applied for detailed investigations of the past lake water column properties (**Figure 5**).

6.2 Chemical composition of the borehole (swamp) section (BS)

Fluctuated values between average to higher values of Si, TiO_2, and Zr elements are observed in sediments retrieved from the swamp section. These elements exhibit similar patterns along this section. Contrarily, CaO, MnO, FeO, and Rb concentrations reflect average to relatively low concentrations. Uppermost two units (Us I and Us II) contain relatively high CaO, K_2O, and MnO contents but low to average concentrations of Si, Ti, Fe_2O_3, and Zr (**Figure 6**). The highest concentrations of Si, Zr, and TiO_2 are observed in unit Us III, which has overall relatively low concentrations of CaO, MnO, FeO, and K_2O. The lowermost unit (Us IV) also reflects a characteristic geochemical signature, with the uppermost part of this unit containing relatively high K_2O, CaO, MnO, FeO, Zn, and Rb and low Si, TiO_2, and Zr concentrations. The general downcore increasing tendency of Si, TiO_2, and Zr concentrations continues in this unit. The rapid variability of the redox-sensitive elements (Zn, V) and V/V + Ni was observed for 2.6–4.7 m interval. Lower values of these elements and element ratios were observed for the 7.9–9.7 and 11.3–12 m intervals and at 6.2 m depth (**Figure 6**). In accordance to V/Cr, U/Th, Ni/Co, and V/V + Ni cross-correlations, BS sediments are placed at the same specific area, which is also characteristic for the BAF37 sediments (**Figure 5**).

7. Discussion

A quantitative chemostratigraphic approach, together with sedimentological observations, has been used for a better understanding of the Late Holocene paleo-ecological history of the Lake Bafa and related aquatic environments. Furthermore, a sum of the selected paleo-ecosystem parameters (e.g., element concentrations and

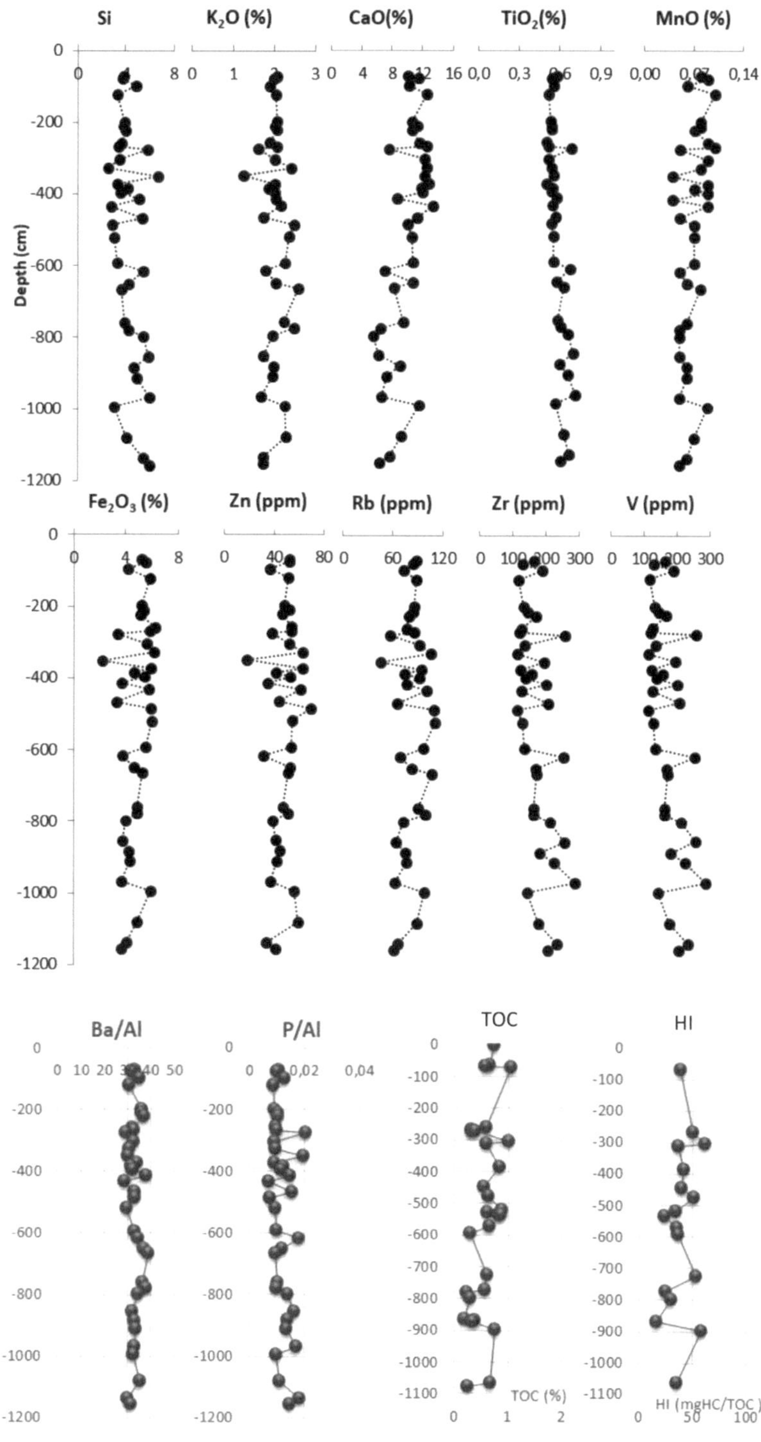

Figure 6.
Concentrations of selected elements and element oxides along the drill core BS, including TOC and HI data, and specific element ratios as a measure for organic matter accumulation rates.

element ratios) have been applied to identify the physical dynamics of the environment and the chemical characteristics of the water column, in terms of oxidation level and salinity. Furthermore, main environmental controls on Lake Bafa aquatic ecosystem were constructed using statistical approaches (FA) on geochemical data.

These controls include the external processes of terrigenous supply, leading to the enrichment of the siliciclastic elements, such as Si, Ti, and K and internal processes of biologic activity, which results in enrichment of Ca, Mg, Sr (in endogenic carbonate), TOC, and P. Herewith, high negative Factor I loads indicate similar geological sources for Al, K, Ti, Fe, Mg, Ni, Ba, and V enrichments for lacustrine sediments (cores BAF37 and BAF3B) (**Table 2**). Taking into consideration the Factors I and II loadings together, two main subgroups are suggested for these sediments. The first group is mainly a clastic-sourced element group (Al, Ti, Fe, Mg, K). The second group is related to the endogenic processes of carbonate deposition (Ca, Sr) and organic productivity (P) and water and sediment column redox processes (Ni, Co, V). The enrichment of these transition metals (Mn, Fe, Ni) is probably controlled by both detrital input and redox processes in the water and sediment columns. The factor load signature indicates that the swamp section reflects similar sources and modes of enrichment for most of these elements (**Table 2**). This suggests that the same geological processes prevailed also in the adjacent swamp area. However, there are also differences; Ba, Sr, and Ti enrichment pathways are different in the swamp sediments. Moreover, the nutrient elements (mainly P) and endogenic carbonate group elements (Ca and Sr) have similar factor (high Factor II and lower Factor I) loadings, suggesting a strong association with the organic productivity.

7.1 Detrital input and changes in hydrological conditions

The interpretation of the detrital sources and transport intensity of the detrital material allows us to determine the physical dynamics and energy conditions of the Lake Bafa Basin. Basically, elemental enrichment of Si, K, Ti, Zr, and Rb indicates the terrigenous material supply. Contrarily Sr, Ca, Mg, and Ba reflect biogenic sources [11, 27–29]. K, Si, and Ti element enrichments and enhanced average grain size distribution of the lake sediments (BAF37), "accumulated during the period of 2.5–2.2 ka year BP," indicate deposition under high energy conditions (**Figure 4**). This likely corresponds to a period when intensive freshwater input occurred during the earlier stages of separation of Lake Bafa from the Aegean Sea. However, this input was interrupted abruptly and followed by a short-term low energy conditions, indicated by the increasing clay size fraction and Ca (carbonate) contents of the sediments [8, 13, 31]. Average detrital input was low during the period of 1.95–0.8 cal. ka year BP, except for a brief period around 1.8 cal. ka year BP when relatively high energy conditions prevailed, which was likely caused by an abrupt hydrological change. After 0.8 cal. ka BP, the lake became completely isolated from the sea [8, 9] but continued to be influenced by the water and sediment inputs in its western part from the Büyük Menderes River. This river flooding events strongly influenced the Lake Bafa hydrology and caused abrupt water level fluctuations as well as rapid increases in the sedimentation rates during the Late Holocene [8, 9, 13, 30, 31]. The high variability of the river discharge was mainly controlled by the climate-driven rainfall pattern. During the enhanced discharge events, transported sediments could easily reach the lake since the streambed slopes have lower gradients. Therefore, we would suggest the climate-driven processes mainly controlled the variability of the hydrological conditions in the isolated lake during the last 800 or so years. Starting from the 1990s, a rubber dam was constructed. Since then, the lake level is currently artificially controlled.

7.2 Organic matter productivity

Relative variations of the biological production rates were determined, applying organic matter accumulation indicators (i.e., P concentrations, Ba enrichments,

TOC concentrations). Phosphorus is known as the main nutrient marker, which would limit the biological productivity in any aquatic ecosystem [27, 32]. Ba enrichments are usually considered as a productivity proxy, being incorporated in the diatom frustules as micro-barite crystals [11, 27, 28]. However, these two elements would easily mobilize in anoxic conditions [33, 34]. Therefore, their signals may partly diminish from the sediments [33, 34]. Particularly, bulk organic carbon content (TOC) indicates changes in accumulation and/or preservation intensity under changing paleoecological conditions [10, 35–37]. Furthermore, hydrogen index (HI) values were applied to determine sedimentary organic matter sources [37, 38].

TOC content of lacustrine sediments is observed in a wide range between 0.7 and 4.4% (BAF37). TOC concentrations of swamp section sediments are quite low (≤1%) (**Figure 6**). Mean TOC content is 2.15% in the lake section and, however, 0.6% in swamp section. This obvious difference probably arises from overall distinct organic matter production processes, accumulation ranges, and preservation conditions. Main nutrient input parameter, P, is relatively low in the older lacustrine sediments (core BAF 37; >1.2 cal ka years). During the recent period, P availability was higher than the mean values. Therefore, P was probably not a limiting factor for organic matter productivity in Lake Bafa water column. Overall Ba trend which is similar to that of P, without obvious fluctuations, supports this conclusion (**Figure 4**).

Detailed investigations of lacustrine sediments supply additional information about time-dependent organic matter accumulation rate signs. Therefore, organic matter transportation and production ranges were probably intensive during the accumulation stages of the older sediments (Units III, IV, and V). During this stage, the organic matter type was probably controlled by a mixed contribution of the aquatic and terrestrial sources, reflected by the relatively higher HI values. Only exceptions are observed in two (2.74 and 3.98 cm) thin layers. These layers have lower HI values, mainly suggesting terrestrial organic matter sources. Similarly, organic matter accumulation of the recent lacustrine sediments (Units I–II; last 0.8-ka-year record) and the entire swamp section (BS) are also sourced by the terrestrial vegetation supply (**Figure 7**).

7.3 Water chemistry

Chemical properties of the water column are evaluated focusing on respective changes in salinity and redox conditions. Mg/Ca provides important evidence for the past salinity history of the water column. Basically, enhanced salinity (Mg/Ca) ratio favors the formation of aragonite, whereas low salinity (Mg/Ca) ratios promote calcite precipitation. This variation has important effects on the biota and the type of carbonate mineral accumulated in the sediments [29, 39]. Minerals containing Ti are conservative in geochemical reactions (e.g., redox conditions) [40]. Contrarily, changing redox conditions strongly affect iron (Fe) and manganese (Mn) cycles either in sediments or water column (trend toward lower pE). In particularly, Mn easily mobilizes in changing redox conditions than Fe [11, 41]. Accordantly, low Mn/Fe ratios reflect reducing conditions. Following the similar aspects V/Cr, U/Th, Ni/Co, U/Th, and V/V + Ni, cross-correlations were used to interpret past redox conditions [11, 12, 42, 43].

The Mg/Ca trend in Lake Bafa's sedimentary successions indicates an upward decrease, which suggests a decrease in salinity in time (**Figure 7**). Time-dependent decrease of the lake water column chemistry from past to recent terms is also supported by the diatom analysis results of the same sedimentary section (core BAF37), published by Bulkan et al. [8, 9].

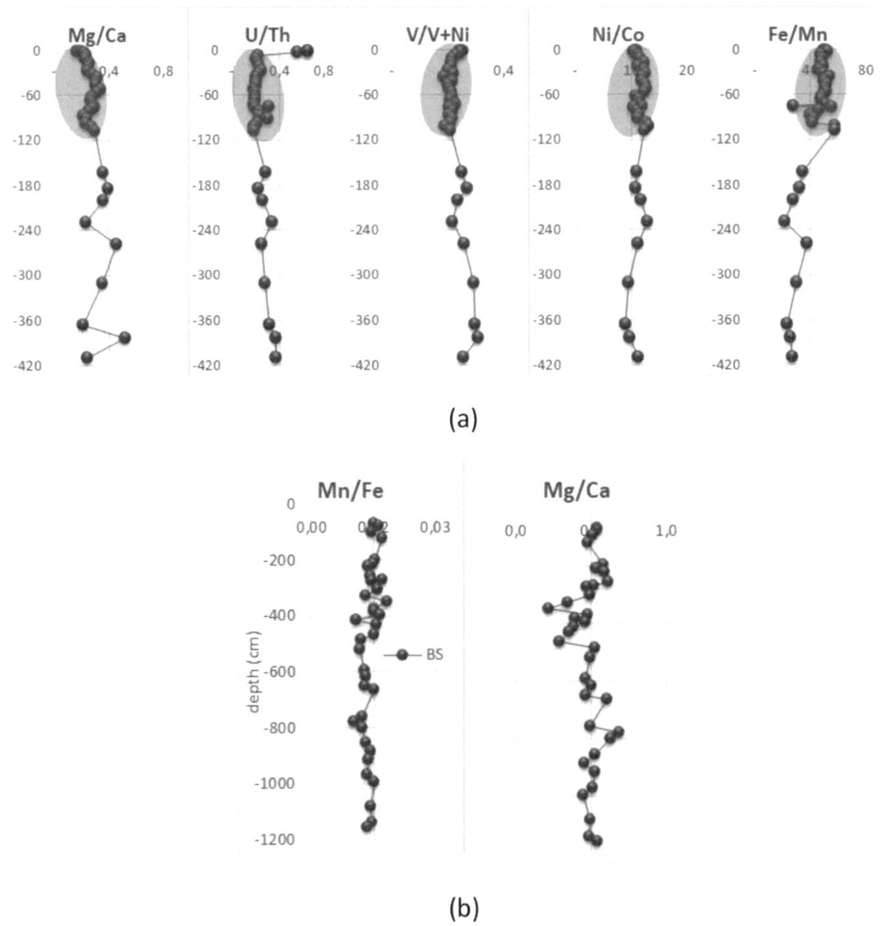

Figure 7.
Element ratio indicator for water chemistry conditions. (a) represents specific element ratios for core BAF37 (blue-colored area represents values measured for sediment–water interface phase and long core sediments, jointly). (b) indicates drill core BS sediments.

Previous studies suggest that Cr, U, and V elements are rare in denitrifying conditions. Contrarily, enhanced Ni, Co, Cu, Zn, Cd, and Mo contributions are documented for organic-rich sediments, accumulated under sulfate-reducing bottom water conditions [11, 44–46]. Cross-correlations of V/Cr–U/Th, Ni/Co–U/Th, V/V + Ni–U/Th, and V/Cr–Ni/Co indicate that both the lake and the swamp sections deposited mainly under oxic conditions (**Figures 5** and 7). Cr, U, and V element ratios in the Lake Bafa core and the swamp section indicate deposition under mainly oxic conditions. Despite most of the redox proxies showing the general oxic water column conditions, Ni/Co ratio points to somehow anoxic conditions. Furthermore, relatively low Mn/Fe ratio of the lake sediments (core BAF-37) indicates that the water column was probably partly oxygen depleted during 1.7–1.4 cal. ka year BP and during the isolated lake period, starting ca. 800 years ago.

7.4 Implications for Holocene age local environmental conditions

Both tectonics and sedimentological processes (N-S extension, graben, horst, erosion, delta progradation) and climate events (both orbital and abrupt changes) have played spectacular roles in modulating the ecological conditions in the Eastern Mediterranean Sea coastal areas. The ecological systems in this region include a variety of terrestrial aquatic ecosystems, ranging from fluvial, lacustrine, and lagoonal to coastal marine environments [6, 47–55]. Additional factors affecting the environmental conditions in such systems include abrupt geological events (e.g., storms, earthquakes, and tsunamis) and global fluctuations of sea level changes related to orbital climatic oscillations. We applied both sedimentological and geochemical proxies to unravel the time-dependent stages of landscape evolution in the coastal area of the Büyük Menderes Graben.

Several geological and geomorphological studies have been carried out in and around the Lake Bafa These studies have contributed to our knowledge of the coastal geomorphological history, sea level changes, tectonic processes driven erosion rates, climate and environmental induced changes (vegetation and faunal changes) and human impact [6, 8, 9, 48–57].

One study by Bruckner et al. [48] suggests that horst and graben tectonics in the Büyük and Küçük Menderes Grabens caused serious environmental and coastline changes. Mullenhoff et al. [6] concluded that the Lake Bafa was formed as a consequence of erosional processes in the adjacent mainland of Turkey, which have controlled the deltaic progradation and the filling of the Latmian Gulf. Particularly, in Miletus and the Büyük Menderes Graben, remarkable transformations have been revealed, with the metamorphosis of the marine gulf into residual lakes (e.g., lakes Azap and Bafa) [48]. The coastline of the Büyük Menderes Delta, located close to the southern graben area, progradaded seaward some 5 km [5] during the last 1.5 ka cal. year BP. Further progradation closed the entrance of the Latmian Gulf that resulted in the isolation of the Lake Bafa from the Aegean Sea at 0.8 cal. ka year BP. Lake Marmara in the Gediz Graben in the north, with a similar setting to that of the Lake Bafa, shows resembling environmental evolution [5, 7, 47], with the detrital input records reflecting synchronous changes within its catchment area. Similar to the Lake Bafa, Marmara witnessed a marked environmental change with a shift toward more oxic and freshwater conditions at 0.95 cal ka year BP and the increased effects of fluvial activity during the last 300 years. [47].

The strongest human impact has been detected at the time of the Greek period in the seventh- to first-century BC and especially during the Roman period in the first-century BC until the fourth-century AD, when sedimentation was about five times higher than that in the periods before and after [22]. These changes were accompanied by changes in the vegetation type. The palynological analyses have shown high amounts of Quercus type before the period of the strong human impact, which is also reflected by the element composition of the relatively recent sediments, water column samples, plant, and shells that have been analyzed for the pollution assessment studies [49–52].

A sum of local environmental conditions in the Lake Bafa and its catchment changed during the last 4.5 cal. ka years, but still several questions in this context remain to be answered. Basic questions are as follows: (I) Are the sediment accumulation rates different during the lacustrine, lagoon, or shallow water environmental phases and transitional stages? [53, 54], (II) how does the radiocarbon reservoir effect varies with changing environmental stages and time? [53, 54], (III) what is the temporal evolution of meandering delta progradation? and (IV) would it be possible to estimate the behavior and frequency of the river flood intensity from the lake's sedimentary record? These questions can be addressed by acquiring and analyzing additional long cores from different parts of the lake and from the delta. Furthermore, the same cores can be used to study rapid geological and climatic events such as tsunamis, volcanic explosions, and storms. Additional methods, including diatom analysis [8, 9], isotope analysis [8, 9], and lipid or amino acid biomarker analyses, would contribute further to our understanding of the ecological evolution of this unique basin.

8. Conclusions

Paleolimnological study of the Middle to Late Holocene lacustrine archive of Lake Bafa allowed us to reconstruct the geological history of the Büyük Menderes Graben's coastal area.

Lithological and geochemical analyses of the 4.2-m-long lacustrine core BAF37 and 12-m-long swamp drill core BS in the adjacent area provide a continuous archive of environmental changes during the last 2.5 and 4.5 cal. ka years, respectively. Geochemical and sedimentological proxy records from both cores show the following significant changes in the ecosystem of the area, from oldest to the present-day: stage IV, marine-dominated (4.5–2.7 cal. ka year BP); stage III, marine–river interaction (2.7–1.75 cal. ka year BP); stage II, lagoon (1.75–0.8 ka year BP); and stage I, the recent isolated swamp lake (last 0.8 cal. ka). This transition from marine stage to the recent lacustrine conditions was somewhat a gradual process, and the environmental conditions during the stage III were predominantly controlled by the tectonics and postglacial sea level rise, resulting in subsidence, erosion, sediment transport, and delta progradation.

The redox conditions were oxic before the isolation but became relatively reducing during the ensuing period. K, Si, and Ti concentrations and grain size distributions reflect a high energy environment interrupted by fluctuating short-term low energy conditions during 2.5–2.2 cal. ka year BP. Low energy conditions also prevailed during the 2.2–0.8 cal. ka yr. BP and particularly during the last 800 years.

Our observations indicate that ecosystem characteristics of the study are controlled by the combination of the hydroclimate and geotectonic processes. However, their effect of intensity is also reflecting changes from past to previous terms.

Acknowledgements

The scientific and technological research council of Turkey (Tubitak-Ardep project numbers: 113Y070; 115Y766) and scientific research projects' coordination unit of Istanbul University (BAP project numbers: 28942; 17828) kindly provided financial research supports, which are gratefully acknowledged. Furthermore, related MSc study project is also supported by scientific research projects' coordination unit of Istanbul University (BAP project number: 37190). We would kindly thank Istanbul Technical University Eastern Mediterranean Centre for Oceanography and Limnology (EMCOL) for their contribution and support for several aspects, i.e., retrieving the core samples, laboratory facilities, and discrete sampling processes. In this respect, we thank EMCOL students especially Dursun Acar, Burak Yalamaz, and Cansu Demirel.

Author details

Özlem Bulkan[1,2]*, Bilgehan Toksoy Ediş[1] and M. Namık Çağatay[3]

1 Geological Engineering Department, Istanbul University-Cerrahpasa, Istanbul, Turkey

2 Department of Surface Waters-Research and Management, Swiss Federal Institute of Aquatic Science and Technology, Kastanienbaum, Switzerland

3 EMCOL Research Centre, Faculty of Mining, Istanbul Technical University, Istanbul, Turkey

*Address all correspondence to: bulkan@istanbul.edu.tr

References

[1] Netto SA, Fonseca G. Regime shifts in coastal lagoons: Evidence from free-living marine nematodes. PLoS One. 2017;**12**(2):1-19. DOI: 10.1371/journal.pone.0172366

[2] Basset A, Sabetta L, Fonnesu A, et al. Typology in Mediterranean transitional waters: New challenges and perspectives. Aquatic Conservation: Marine and Freshwater Ecosystems. 2006;**16**:441-455. DOI: 10.1002/aqc.767

[3] Taymaz T, Yilmaz Y, Dilek Y. The geodynamics of the Aegean and Anatolia: Introduction. Geological Society of London, Special Publication. 2007;**291**(1):1-16. DOI: 10.1144/sp291.1

[4] Schilman B, Bar-Matthews M, Almogi-Labin A, Luz B. Global climate instability rejected by Eastern Mediterranean marine records during the late Holocene. Palaeogeography, Palaeoclimatology, Palaeoecology. 2001; **176**:157-176

[5] Hakyemez HY, Tevfik E, Göktaş F. Late Quaternary evolution of the Gediz and Büyük Menderes grabens, Western Anatolia, Turkey. Quaternary Science Reviews. 1999;**18**:549-554

[6] Müllenhoff M, Handl M, Knipping M, Brückner H. The evolution of Lake Bafa (Western Turkey)-sedimentological, microfaunal and palynological results. Coastline Reports. 2004;**1**:55-66

[7] Brückner H, Müllenhoff M, Handl M, van der Borg K. Holocene landscape evolution of the Büyük Menderes alluvial plain in the environs of Myous and Priene (western Anatolia, Turkey). Zeitschrift für Geomorphologie. 2002; **127**:47-65

[8] Bulkan Ö, Sancar Ü, Sarı E, Kırcı-Elmas E, Yıldız T, Çağatay MN, Acıpınar S, Demirel C, Toksoy B, Acar D. Bafa Gölündeki Holosen İklimsel Değişimlerinin Organik Maddenin Karbon-Hidrojen İzotopik Bileşimi, Lipid Biyolojik İşaretçiler (Uzun Zincirli Alkenon: Lcas) ve Diyatome Kayıtlarına Yansımaları, Final Report, Tubitak, P. Nr.113Y070 (unpublished, in Turkish), 2018. p. 265

[9] Bulkan Ö, Sancar Ü, Wei W, Zhu X, Çağatay MN. Water column chemistry of Late Holocene Lake Bafa, Eastern Coast of the Aegean Sea (Turkey). In: Advances in Science, Technology & Innovation Patterns and Mechanisms of Climate: Paleoclimate and Paleo-environmental Changes from Low-Latitude Regions. Switzerland: Springer Nature Pub.; 2019. pp. 65-67. DOI: 10.1007/978-364030-01599-2_1

[10] Meyers PA. Applications of organic geochemistry to paleolimnological reconstructions: A summary of examples from the Laurentian Great Lakes. Organic Geochemistry. 2003;**34**: 261-289

[11] Riquier L, Tribovillard N, Averbuch O, Devleeschouwer X, Riboulleau A. The Late Frasnian Kellwasser horizons of the Harz Mountains (Germany): Two oxygen-deficient periods resulting from different mechanisms. Chemical Geology. 2006;**233**(1–2):137-155. DOI: 10.1016/J.CHEMGEO. 2006.02.021

[12] Racki G, Racka M, Matyja H, Devleeschouwer X. The Frasnian/Famennian boundary interval in the South Polish–Moravian shelf basins: Integrated event-stratigraphical approach. Palaeogeography Palaeoclimatology Palaeoecology. 2002; **181**(1–3):251-297. DOI: 10.1016/S0031-0182(01)00481-3

[13] Bulkan Ö, Erken Ö, Can H, Yalamaz B. Bafa Gölü Sedimanlarında Kütle

Akması Birimlerinin Tespiti Ve Son 1500 Yılda Gelişmiş Ani Jeolojik Olayların Yorumlanması. Final Report, Tubitak, P.Nr.115Y766 (unpublished, in Turkish). 2017

[14] Reimer PJ, Bard E, Bayliss A, Beck JW, Blackwell PG, Bronk Ramsey C, et al. IntCal13 and Marine13 radiocarbon age calibration curves 0–50,000 years cal BP. Radiocarbon. 2013;**55**(4):18

[15] Talma AS, Vogel JC. A Simplified Approach to Calibrating C14 Dates. Radiocarbon. 1993;**35**(2):317-322

[16] Philippson A. Reisen Und Forschungen Im Westlichen Kleinasien. Gotha: Justus Perthes; 1913

[17] McKenzie D. Active tectonics of the Mediterranean regions. Geophysical Journal of the Royal Astronomical Society. 1972;**30**:109-185

[18] Bozkurt E, Oberhansli R. Menderes Massif (western Turkey): Structural, metamorphic and magmatic evolution. A synthesis. International Journal of Earth Sciences. 2001;**89**:679-708

[19] Kazancı N, Dündar S, Alçiçek MC, Gürbüz A. Quaternary deposits of the Büyük Menderes Graben in western Anatolia, Turkey: Implications for river capture and the longest Holocene estuary in the Aegean Sea. Marine Geology. 2009;**264**(3-4):165-176. DOI: 10.1016/J.MARGEO.2009.05.003

[20] Yılmaz Y, Genç Ş, Gürer F, Bozcu M, Yılmaz K, Karacık Z, et al. When did the western Anatolian grabens begin develop. The Geological Society of London. 2000;**173**(1):353-384

[21] Bozkurt E. Timing of extension on the Büyük Menderes Graben, western Turkey, and its tectonic implications. In: Bozkurt E, Winchester JA, Piper JDA, editors. Tectonics and Magmatism in Turkey and the Surrounding Area. Vol. 173. London: Geological Society, Special Publications; 2000. pp. 385-405

[22] Knipping M, Müllenhoff M, Brückner H. Human induced landscape changes around Lake Bafa (western Turkey). Vegetation History and Archaeobotany. 2008;**17**:365-380

[23] Sarı HM, Balık S. Bafa Gölündeki Ceran Balığı (*Liza ramada* Risso, 1826) Populasyonunun Biyolojik Yönden İncelenmesi. Su Ürünleri Dergisi (In Turkish). 199; **613**(3-4): pp. 305-316

[24] Erdogan S. A chemical reaction to a physical impact: Lake Bafa wetland ecosystem (Turkey) case. Ankara Üniversitesi Cevrebilimleri Dergisi (In Turkish). 2011;**3**(1):1-8

[25] Magnin G, Yarar M. Important Bird Areas in Turkey. İstanbul, Turkey: Doğal Hayatı Koruma Derneği; 1997. pp. 78-80

[26] Julia R, Loque JA. Climatic changes vs. catastrophic events in lacustrine systems: A geochemical approach. Quaternary International. 2006;**158**: 162-171

[27] Pujol F, Berner Z, Stüben D. Palaeoenvironmental changes at the Frasnian/Famennian boundary in key European sections: Chemostratigraphic constraints. Palaeogeography, Palaeoclimatology, Palaeoecology. 2006; **240**:120-145

[28] Tribovillard N, Algeo TJ, Lyons T, Riboulleau A. Trace metals as paleoredox and paleoproductivity proxies: An update. Chemical Geology. 2006;**232**:12-32

[29] Dean JP, Bradbury WE. Evidence for Rapid Climate Change in the North-Central United States. Elk Lake, Minnesota: Geological Society of America; 1993. p. 336

[30] Durdu ÖF. Effects of climate change on water resources of the Büyük Menderes river basin, Western Turkey. Turkish Journal of Agriculture and Forestry. 2010;**34**:319-332. DOI: 10.3906/tar-0909-402

[31] Bulkan Ö. Events and cycles: Great Menderes Floodplain (Western Anatolia, Turkey). Nagoya: XIXth INQUA; 2015

[32] Schmitz B, Charisi SD, Thompson EI, Speijer RP. Barium, SiO_2 (excess), and P_2O_5 as proxies of biological productivity in the Middle East during the Palaeocene and the latest Palaeocene benthic extinction event. Terra Nova. 1997;**9**:95-99

[33] Ingall E, Jahnke R. Influence of water-column anoxia on the elemental fractionation of carbon and phosphorus during diagenesis. Marine Geology. 1997;**139**:219-229

[34] McManus J, Berelson WM, Klinkhammer GP, Johnson KS, Coale KH, Anderson RF, et al. Geochemistry of barium in marine sediments: Implications for its use as a paleoproxy. Geochimica et Cosmochimica Acta. 1998, 1998;**62**:3453-3473

[35] Ariztegui D, Chondrogianni C, Lami A, Guilizzoni P, Lafargue E. Lacustrine organic matter and the Holocene paleoenvironmental record of Lake Albano (Central Italy). Journal of Paleolimnology. 2001;**26**(3):283-292

[36] Ariztegui D, Farrimond P, McKenzie JA. Compositional variations in sedimentary lacustrine organic matter and their implications for high Alpine Holocene environmental changes: Lake St. Moritz, Switzerland. Organic Geochemistry. 1996;**24**:453-461

[37] Espitalie J, Laporte JL, Madec M, Marquis F, Leplat P, Paauulet J, et al. Méthode rapide de caractérisation des roches mères, de leur potentiel pétrolier et de leur degré d'évolution. Revue de l'Institut Francais du Petrole. 1977;**32**: 23-42

[38] Talbot MR, Livingstone DA. Hydrogen index and carbon isotopes of lacustrine organic matter as lake level indicators. Paleogeography, Paleoclimatology, Paleoecology. 1999; **70**:121-137

[39] Müller G, Irion G, Förstner U. Formation and diagenesis of inorganic Ca–Mg carbonates in the lacustrine environment. Naturwissenschaften. 1972;**59**:158-164

[40] Boyle JF. Rapid elemental analysis of sediment samples by isotope source XRF. Journal of Paleolimnology. 2000; **23**:213-221

[41] Calvert SE, Pedersen TF. Sedimentary geochemistry of manganese: Implications for the environment of formation of manganiferous black shales. Economic Geology. 1996;**91**:36-47

[42] Jones B, Manning DAC. Comparison of geochemical indices used for the interpretation of paleoredox conditions in ancient mudstones. Chemical Geology. 1994;**114**:111-129

[43] Rimmer SM. Geochemical paleoredox indicators in Devonian–Mississippian black shales, Central Appalachian Basin (USA). Chemical Geology. 2004;**206**(3–4):373-391. DOI: 10.1016/J.CHEMGEO.2003. 12.029

[44] Huerta-Diaz MA, Morse JW. A quantitative method for determination of trace metal concentrations in sedimentary pyrite. Marine Chemistry. 1990;**29**:119-144. DOI: 10.1016/ 0304-4203(90)90009-2

[45] Huerta-Diaz MA, Morse JW. Pyritization of trace metals in anoxic marine sediments. Geochimica et Cosmochimica Acta. 1992;**56**(7):

2681-2702. DOI: 10.1016/0016-7037(92)90353-K

[46] Morse JW, Luther GW. Chemical influences on trace metal-sulfide interactions in anoxic sediments. Geochimica et Cosmochimica Acta. 1999;**63**:3373-3378. DOI: 10.1016/S0016-7037(99)00258-6

[47] Bulkan Ö, Wilkes H, Yalçın MN. Geochemistry of Marmara Lake sediments implications for Holocene environmental changes in Western Turkey. Quaternary International. 2018;**486**:199-214. DOI: 10.1016/j.quaint.2017.12.045

[48] Brückner H, Herdl A, Müllenhoff M, Stock F. Life cycle of estuarine islands- From the formation to the landlocking of former islands in the environs of Miletos and Ephesos in western Asia Minor (Turkey). Journal of Archaeological Science: Reports. 2017;**12**:876-894

[49] Yilgör S, Kucuksezgin F, Ozel E. Assessment of metal concentrations in sediments from Lake Bafa (Western Anatolia): An index analysis approach. Bulletin of Environmental Contamination and Toxicology. 2012;**89**(3):512-518

[50] Bulkan Ö. Holocene Marine-Lake transition effect on ecosystem and local climate around Lake Bafa and Karine Lagoon (Aydın) Final Report: Istanbul University Research Fund P. Nr: 28942 (unpublished, in Turkish). 2014

[51] Bulkan Ö, Acıpınar B. Plant leave, sediment and soil chemistry: An ecosystem model for the surroundings of the Lake Bafa (Western Anatolia). International Meeting on Organic Geochemistry. 2015;**1**:668

[52] Sarı E, Bulkan Ö, Çağatay MN, Toksoy B, Kurt MA. Ecological risk assessment of heavy metals in core sediments and mussel samples from the Lake Bafa, Turkey. 30th Society of Environmental Geochemistry and Health International Conference (SEGH). 2014. pp. 30-31

[53] Thonemann P. The Maeander Valley: A historical geography from Antiquity to Byzantium. United Kingdom: Cambridge University Press; 2011

[54] Izdebski A. Environmental history of the hinterland. In: Niewöhner P et al., editors. The Byzantine history of Miletus and its hinterland—Quantitative Aspects: Stratigraphy, Pottery, Anthropology, Coins, and Palynology. Vol. 2. Berlin: Archäologischer Anzeiger; 2016. pp. 270-280

[55] Bulkan Ö. As a pilot study: Organic geochemical characterization of climate sensitive plants around Istanbul-Kocaeli area. Istanbul University Research Fund P Nr: 17828 (unpublished, in Turkish). 2013

[56] Acıpınar S. Bafa gölüne ait genç çökellerin ve çevre florasinin karbon, hidrojen, azot izotopik bileşimi ve Holosen ekosistemi. Istanbul University MSC thesis, (unpublished, in Turkish). 2017

[57] Toksoy B. Bafa gölü çökellerinin lito-biyo-kemo stratigrafik özellikleri ve kuvaterner ekosistemi, Istanbul University MSC thesis, (unpublished, in Turkish). 2013

Chapter 4

Fingerprinting Sources of the Sediments Deposited in the Riparian Zone of the Ruxi Tributary Channel of the Three Gorges Reservoir (China)

Zhonglin Shi, Dongchun Yan, Anbang Wen and Yongyan Wang

Abstract

The riparian zone of the Three Gorges Reservoir serves as a critical transitional zone located between the aquatic and surrounding terrestrial environments. The periodic anti-seasonal alternation of wet and dry periods results in an intensive exchange of substance within the riparian zone. The discrimination of the sources of the sediments deposited within the riparian zone is of fundamental importance for the evaluation of the soil pollution and associated environmental impacts and for the protection of the water quality in the reservoir. In this study, a composite fingerprinting technique has been applied to apportion the sediment sources for the riparian zone with different elevations, ranging between 145—155, 155–165, and 165–175 m in a typical tributary channel. From a sediment perspective, the sediments suspended from the Yangtze mainstream represent the primary sources of the riparian deposits. From a contamination perspective, the sediment input from the Ruxi tributary channel represents an important source of pollution for the riparian environment. More effective sediment and sediment-associated contaminant control plans are needed to reduce the potential environmental problems of the riparian zone.

Keywords: riparian zone, sediment source, fingerprinting, contaminant, Three Gorges Reservoir

1. Introduction

Since the full impoundment of the Three Gorges Reservoir in late 2010, a distinctive reservoir marginal landscape has been created, which is commonly known as the riparian zone [1, 2]. This zone, also named as water-level fluctuation zone [3] or littoral zone [4], is a unique artificial landscape that formed after the construction of the Three Gorges Dam on the Yangtze River. Definitely, the reservoir riparian zone refers to the elevation ranges between the base water level of 145 m in wet season for flood control and the peak level of 175 m in dry season for energy generation. Unlike the traditional riparian zones in natural river systems that were affected by irregular or occasional overbank flooding, the reservoir riparian zone is characterized by the regular water level fluctuation as a result of the reservoir impoundment [5].

The riparian zone of the Three Gorges Reservoir has a vertical height of 30 m, and extends ca. 663 km along the mainstream of the Yangtze River, with a total area of 349 km^2 [3]. After its generation, the riparian zone has been subjected to an annually cyclic inundation and exposure. As a critical transitional zone between the aquatic and terrestrial ecosystems, numerous environmental issues related to this zone have captured extensive attention in recent years, such as bank erosion [6], revegetation [7], sedimentation, and associated contamination [1, 2, 8].

During the impounding period of the reservoir, which typically extends from September to next June, the increased water depth and decreased flow velocity provide much opportunities for the deposition and storage of sediment and sediment-associated contaminants, both within the channel and on the areas bordering the channel, for example, the riparian zone. Despite no obvious deterioration in water quality in the mainstream of the Yangtze River, most of the tributaries have experienced serious eutrophication since the operation of the reservoir [9]. Serving as an important carrier of contaminants, the sediment deposited and subsequently stored within the riparian zone of the reservoir may also result in the pollution of the environment. Within the exposure period, which represents the wet season of the Yangtze River basin, remobilization of the contaminated sediment deposited within the riparian zones by slope and bank erosions may reintroduce nutrients and contaminants into the watercourse. Moreover, the repeated agricultural use of the riparian areas will have the potential to transmit the enriched contaminants to human via food chains. In this context, information on the source of the riparian deposits is needed for the development of environmentally sound water and sediment management strategies.

Sediment fingerprinting is a widely employed approach to quantify the relative contribution of potential sources to the target sediment, such as suspended sediments delivered either in riverine systems or at the outlet of a catchment or deposits collected from floodplains, reservoirs, wetlands, and lakes. This technique can be traced back to the 1970s, and the past 40 years has witnessed the progressive development and refinement of the approach [10]. The most important assumption underpinning the fingerprinting technique is that one or more of the physical or chemical properties (i.e., fingerprints) could clearly differentiate potential sources of the sediment. A wide range of soil and sediment properties have now been successfully used as fingerprints, including mineral magnetics, fallout radionuclides, color, geochemistry, and stable isotopes [11]. Along with the increasing number of properties being used in fingerprinting studies, there is a need to select the "best" set of properties to discriminate between sediment sources. Although a lot of statistical analyses have been tested to identify the optimum combination of those properties, one commonly adopted procedure involves a two-step process, which combined Kruskal-Wallis H-test as first step and discrimination function analysis as second step [12]. A recent work has also demonstrated that this two-step fingerprint selection procedure (KW + DFA) was found to be the most effective option, which provides the most reliable source apportionment results in their study catchment [13].

The Ruxi River catchment is located in the middle section of the Three Gorges Reservoir region and drains an area of 721.4 km^2. The river extends approximately 54.5 km and is a first-order tributary of the Yangtze River (**Figure 1**). This catchment is characterized by a subtropical humid monsoonal climate. The mean annual rainfall is about 1140 mm, with approximately 70% falls between May and September. The soils are purple soils and the land uses are dominated by arable land

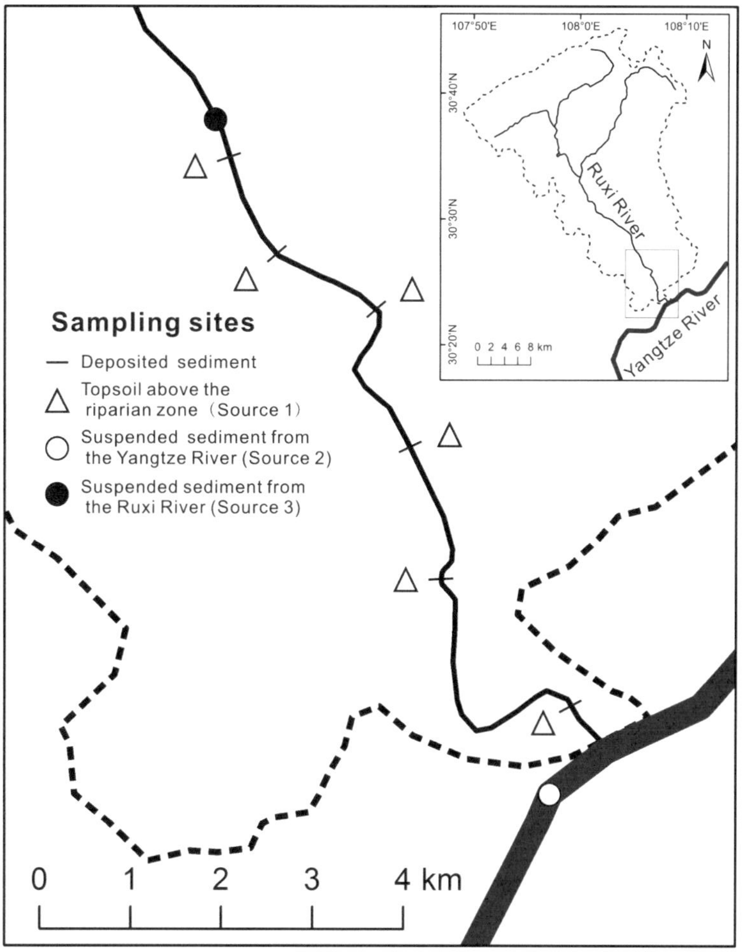

Figure 1.
The study area and sampling sites.

and woodland. The riparian zone in the Ruxi catchment covers an area of 6.4 km^2 and extends about 6 km upstream from its confluence with the Yangtze mainstream. The riparian zone is characterized by gentle slopes with the gradient less than 15° and is subjected to evident sediment deposition during the impounding period of the reservoir.

In this study, the sediment source fingerprinting technique was applied to discriminate deposited sediment sources in the riparian zone of the Ruxi tributary channel in the Three Gorges Reservoir, China. The main objectives were (i) to test the feasibility of using fingerprinting approach to estimate the relative contribution of potential sources to the sediment deposited within the riparian zone and (ii) to explore the environmental implications of sediment sources.

2. Materials and methods

2.1 Sediment source classification

The identification and the classification of the potential sediment sources within a catchment or river basin is of fundamental importance for a successful application of the fingerprinting approaches. The sources have been classified based on the source type (e.g., surface or sub-surface erosion, areas under different land use) or on spatial location (e.g., sub-catchments or geological units). Despite that most existing studies have focused on discriminating sources types [10], information on spatial source, for example tributary sub-basins, might be more important at larger scales [14].

Three spatially differentiated sources have been identified in this study to apportion their contribution to the sediments deposited within the riparian zone of the Ruxi tributary channel. Firstly, the upside soils above the 175 m elevation level have the possibility to be moved downward as a result of surface erosion associated with rainfall events and agricultural disturbance, which therefore provide a potential source (Source 1) of the sediment. Secondly, the suspended sediments transported by both the mainstream of Yangtze River (Source 2) and the upstream Ruxi River (Source 3) are also likely to be deposited and stored within the riparian zone during the impounding period of the reservoir.

2.2 Sample collection

Representative samples of the deposited sediments have been collected from the riparian zone at six sections along the Ruxi River by using artificial grass mats as sediment traps. At each section, the 30 m high riparian zone has been subdivided into three intervals of elevation (145–155, 155–165, and 165–175 m). For each interval, one small piece of plastic mat with an area of 1 × 1 m has been fixed to the soil surface with steel pins prior to inundation. The time-integrated deposited sediments have been sampled between September 2015 and June 2016, encompassing a complete inundation and exposure period of the riparian zone.

At each section for deposited sediment collection, surface soils (0–2 cm) have been grabbed at locations with actively eroding signs and immediately above the 175 m elevation level. To increase the representativeness of the source samples, several subsamples have been collected within the vicinity of each sampling location and then mixed to produce a composite sample (n = 6).

The collection of fluvial suspended sediments (Source 2 and Source 3) has been undertaken monthly from September 2015 to June 2016. One time-integrating sediment trap has been deployed in the main channel of the Yangtze River, which is located at approximately 1 km upstream from the confluence with the Ruxi River (**Figure 1**). The trap consists of a floating barrel fixed at 1.5 m below the ambient water surface. Using the same method, suspended samples of the Ruxi tributary have been collected from a site near the center of the river channel cross section. This sampling site was located about 5 km upstream of the tributary. Totally, 18 samples of the suspended sediment transported by both the Yangtze mainstream (n = 9) and Ruxi River (n = 9), which representing Source 2 and Source 3, respectively, were collected.

It is noted that, however, the specific flow regulation mode of the Three Gorges Dam leads to remarkable difference in water level residence time and inundation duration between different elevations within the 30-m-height zone. Typically, water level rises rapidly from 145 to 175 m during the period spanning from mid-September to early October; then, it remains at the peak level until late January. Subsequently, water level retreats gradually to the 145 m base level in early June, which then maintain around this level until next inundation period (**Figure 2**). Therefore, deposited sediment sources from the suspended material originating from the mainstream and tributary vary temporally with the variation of water levels. In the case of the 145–155 m elevation, the inundation duration lasts for nearly 9 months. For the upper 155–165 and 165–175 m elevations, the riparian zones are submerged for about 6 and 4 months, respectively. In the case of the

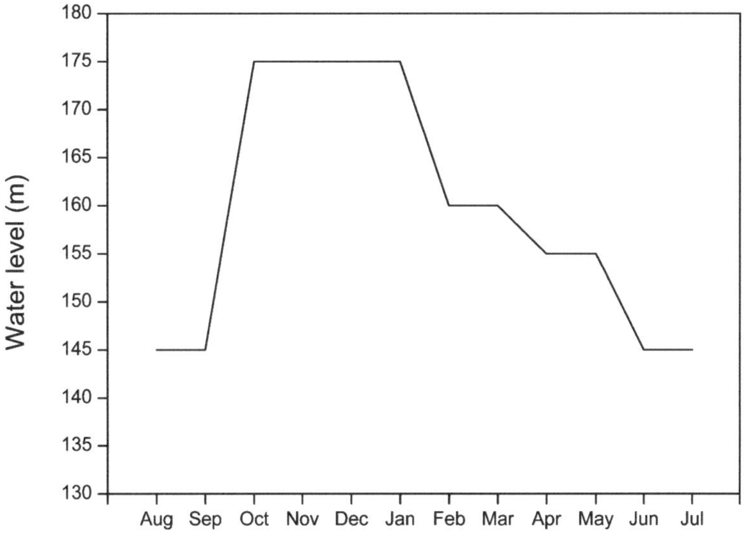

Figure 2.
Dynamic of water levels with the inundation period of the Three Gorges Reservoir.

Source 1, however, the eroded material has the equal chance to be transported downward and stored within different elevations. In this context, sediment sources have been allocated according to the submerging and sampling periods with the exception of Source 1.

2.3 Laboratory analysis

The deposited sediment collected on each mat has been rinsed using deionized water and recovered by sedimentation and centrifugation. After being retrieved from the traps, the suspended sediment samples (Source 2 and Source 3) have been also obtained by setting and centrifugation in the laboratory. In the case of the composite topsoil samples, stones and visible plant debris have been discarded before being dried.

All samples have been air-dried at room temperature, gently disaggregated using a pestle and mortar and screened through a 2-mm sieve. Subsamples of the <2 mm fraction of the target deposit were measured firstly to determine the grain size distribution using a laser diffraction granulometer (Mastersizer 2000, Malvern Instruments). Prior to analysis, the samples were pretreated with 10% H_2O_2 and 10% HCl to remove organic matter and $CaCO_3$, respectively, and then dispersed with ultrasound for 2 minutes. The obtained results have revealed that the deposited sediments collected from the riparian zone with different elevation levels were dominated by silt and clay particles, in which over 90% of the sediment was <63 μm in size. Consequently, all source material and deposited sediment samples were sieved to <63 μm to obtain a comparable grain-size faction between source and sediment material. Subsequent analyses were restricted to the <63 μm fraction.

A total of 18 potential fingerprinting properties were selected for analysis. Concentrations of TOC and TN were measured using a vario MACRO cube element analyzer (Elementar, Germany) after the removal of $CaCO_3$ using 10% HCl. Other elements, including TP, K, Mg, Na, Ca, Fe, Al, Cr, Mn, Ti, Zn, Cd, Co, Cu, Ni, and Pb, were determined using ICP-OES and ICP-MS following a microwave-assisted digestion with HNO_3 and HF. Duplicates, method blanks, and standard reference materials (GBW07401) were used for quality assurance and control.

Property	145–155 m				155–165 m				165–175 m			
	S1	S2	S3	Target	S1	S2	S3	Target	S1	S2	S3	Target
TN	0.725	0.407	0.114	0.430	0.725	0.131	0.009*	0.910	0.725	0.496	0.009*	0.451
TOC	0.180	0.299	0.052	0.432	0.180	0.074	0.006*	0.625	0.180	0.625	0.699	0.825
TP	0.490	0.405	0.656	0.537	0.490	0.723	0.359	0.843	0.490	0.204	0.521	0.747
Fe	0.074	0.994	0.795	0.745	0.074	0.948	0.861	0.470	0.074	0.899	0.648	0.160
Al	0.071	0.549	0.975	0.133	0.071	0.376	0.523	0.910	0.071	0.297	0.923	0.206
Ca	0.114	0.020*	0.003*	0.516	0.114	0.005*	0.001*	0.927	0.114	0.594	0.350	0.722
Mg	0.667	0.987	0.684	0.207	0.667	0.723	0.573	0.388	0.667	0.676	0.616	0.696
K	0.167	0.982	0.978	0.235	0.167	0.758	0.911	0.432	0.167	0.387	0.973	0.042*
Na	0.569	0.006*	0.116	0.396	0.569	0.094	0.272	0.878	0.569	0.817	0.150	0.052
Ti	0.323	0.811	0.052	0.384	0.323	0.861	0.066	0.177	0.323	0.641	0.259	0.574
Mn	0.371	0.951	0.260	0.637	0.371	0.966	0.666	0.000*	0.371	0.993	0.504	0.811
Co	0.694	0.636	0.466	0.726	0.694	0.655	0.256	0.101	0.694	0.326	0.404	0.648
Cu	0.043*	0.247	0.003*	0.253	0.043*	0.507	0.024*	0.806	0.043*	0.131	0.015*	0.027*
Cr	0.999	0.134	0.608	0.945	0.999	0.118	0.339	0.717	0.999	0.077	0.390	0.961
Zn	0.071	0.927	0.566	0.206	0.071	0.020*	0.729	0.936	0.071	0.013*	0.037*	0.867
Pb	0.213	0.221	0.087	0.946	0.213	0.420	0.180	0.608	0.213	0.544	0.738	0.101
Cd	0.021*	0.963	0.800	0.868	0.021*	0.341	0.505	0.688	0.021*	0.152	0.142	0.407
Ni	0.114	0.542	0.561	0.835	0.114	0.626	0.431	0.074	0.114	0.686	0.259	0.748

*Statistically significant values at $P \leq 0.05$.

Table 1.
The results of the Shapiro-Wilk test for normality.

Elevation level (m)	Properties passed the range test
145–155	TN, TOC, TP, Na, Ti, Mn, Cr, Zn, Pb, Cd
155–165	TN, TOC, TP, Ca, Mg, Na, Ti, Mn, Cu, Cd
165–175	TN, TOC, TP, Fe, Al, Ca, Mg, K, Na, Ti, Mn, Cu, Cr, Pb, Cd

Table 2.
Fingerprinting properties passed the range test for different elevation levels in the riparian zone.

145–155 m			155–165 m			165–175 m		
Property	*H*-value	*P*-value	Property	*H*-value	*P*-value	Property	*H*-value	*P*-value
TN	13.814	0.001*	TN	11.275	0.004*	TN	8.857	0.012*
TOC	14.658	0.001*	TOC	11.275	0.004*	TOC	8.857	0.012*
TP	14.487	0.001*	TP	12.329	0.002*	TP	9.736	0.008*
Na	9.937	0.007*	Ca	12.784	0.002*	Fe	4.295	0.117
Ti	16.275	0.000*	Mg	11.368	0.003*	Al	1.929	0.381
Mn	14.656	0.001*	Na	9.697	0.008*	Ca	11.429	0.003*
Cr	5.889	0.053	Ti	11.699	0.003*	Mg	8.000	0.018*
Zn	10.926	0.004*	Mn	12.784	0.002*	K	1.529	0.466
Pb	12.010	0.002*	Cu	3.029	0.220	Na	5.618	0.060
Cd	13.469	0.001*	Cd	12.501	0.002*	Ti	8.324	0.016*
						Mn	11.000	0.004*
						Cu	4.524	0.104
						Cr	4.614	0.100
						Pb	10.629	0.005*
						Cd	10.315	0.006*

*Statistically significant values at $P \leq 0.05$.

Table 3.
The results of applying the Kruskal-Wallis H-test to the fingerprint property dataset for different elevation levels in the riparian zone.

2.4 Sediment source discrimination

The measured values have been firstly tested for normality prior to proceed with the statistical selection of potential fingerprint properties [15]. Particle size and organic matter correction factors were not included in this study to avoid the risk of over-correction of the tracer values [16, 17]. **Table 1** shows that some properties exhibit non-normal distribution. Nonetheless, the majority of the fingerprint properties for source and target samples in different elevation levels have passed the Shapiro-Wilk test, confirming that these data were normal in distribution. Since the number of samples in this study is relatively small and the mean values are more sensitive than the median, the median values for individual properties in both source and target samples are used for further analyses.

One fundamental assumption underpinning the sediment source fingerprinting techniques is that the tracer properties, which were selected to discriminate between sources, should behave conservatively during mobilization and delivery through the catchment system [14]. Although the conservative behavior of the

Elevation level (m)	Step	Fingerprint	Wilks' λ	Cumulative % source type samples classified correctly
145–155	1	Cd	0.367	62.5
	2	Ti	0.138	91.7
	3	Mn	0.088	95.8
155–165	1	Mg	0.277	61.1
	2	Mn	0.075	94.4
	3	TP	0.026	100.0
	4	TOC	0.014	100.0
165–175	1	Ca	0.016	100.0
	2	TN	0.001	100.0
	3	TP	0.001	100.0

Table 4.
The optimum composite fingerprint for different elevation levels in the riparian zone established using stepwise discriminant function analysis.

tracer properties is complex and difficult, if not impossible, to quantify, a modified range test was applied to ensure that the median concentration for each tracer associated with the target falls within the range of median concentrations of that tracer associated with the potential sources [18]. The potential fingerprinting properties, which passed the range test for each elevation level, are listed in **Table 2**. It has been recognized that such a range test is simply a method of selection to exclude the properties suffering a significant change during the transport within the fluvial system; thus, complete absence of tracer property transformation (i.e., conservative behavior) is not guaranteed [19].

A two-stage statistical procedure [12] was then applied to the source material properties passing the range test to identify the composite fingerprints that provide a good discrimination between sources. During the first stage, the Kruskal-Wallis H-test was used to select properties exhibiting significant differences ($P \leq 0.05$) between the individual sources (**Table 3**). Properties passing the Kruskal-Wallis H-test are then tested, in stage two, by stepwise multivariate discriminant function analysis (DFA) to identify the optimum combination of tracers for discriminating the source groups (**Table 4**). In this stage, each fingerprint property was selected based on the minimization of Wilks' λ and a probability value for parameter entry of 0.05.

3. Results

A frequentist-based multivariate mixing model [12] was applied to assess the relative contribution of the three potential sources to the deposited sediment samples collected from each designated riparian zone. In this method, the proportions P contributed by the m individual sources s are quantified by minimizing the sum of the squares of the residuals (R_{es}) for the n tracer properties involved, where:

$$R_{es} = \sum_{i=1}^{n} \left(\frac{C_{di} - \left(\sum_{s=1}^{m} C_{si} P_s\right)}{C_{di}} \right)^2 \tag{1}$$

and C_{di} is the concentration of tracer property i in the deposited sediment sample, C_{si} is the concentration of tracer property i in source group s and P_s is the relative proportion from source group s. Note that two constraints are imposed on the mixing model: (a) the relative contributions from the individual sediment sources must lie between 0 and 100%, and (b) these contributions sum to 100%.

The mixing model was optimized using the OptQuest algorithm in Oracle's Crystal Ball software. To address the uncertainties associated with representation of the sources and targets by single property values (e.g., mean or median), Student's t-distributions were assigned for each source (C_{si}) and target sediment property (C_{di}) using measured median as location and the product of standard deviation and $n^{-1/2}$ as scale parameter, where n is the number of samples [20]. The proportional contribution for each source was repeatedly solved for 1000 times using a Latin Hypercube sampling method.

The results of Latin Hypercube sampling procedure documented that the estimates of sediment contribution to the riparian deposits from the individual sources involved limited uncertainty bands of ±0.5–0.7% at the 95% level of confidence. Consequently, the proportional source contributions are reported as absolute median values in the following section when comparing and explaining the results.

The robustness of the optimized solutions of the mixing model was assessed using a goodness-of-fit (GOF) function, which compares the actual fingerprint property concentration measured in target sediment with the corresponding values predicted by the model, based on the optimized percentage contribution from each source group [21]. The GOF function is defined as:

$$GOF = 1 - \left[\frac{1}{n}\sum_{i=1}^{n}\frac{\left|C_{di} - \sum_{s=1}^{m} C_{si}P_s\right|}{C_{di}}\right] \tag{2}$$

Figure 3 presents the median contributions from individual source types to the deposited sediment samples collected from the riparian zone with different elevation level. The GOF values range between 0.88 and 0.95, indicating that the modeling results are acceptable. The sources of the suspended material from the Yangtze mainstream (Source 2) contributed the most proportion (>40%) to the deposited sediment for all three elevation levels of the riparian zone. There were, however, clear evidence of significant contrasts between the contributions from the other two sources (i.e., sources 1 and 3) for the 145–155 m elevation level and the other two upper elevation levels. In the case of the sources of topsoil above the riparian zone (Source 1), the estimated contribution to the lowest 145–155 m deposits was only about half of those to the other two upper elevation levels. In the case of the sediment input from the Ruxi tributary (Source 3), the relative contribution decreased with increasing elevation, despite that the values were comparable between the upper 155–165 and 165–175 m levels.

4. Discussion

4.1 Sediment source apportionment

A good discrimination between the sources (**Table 3**) and high levels of correct classification of source type samples (**Table 4**) were provided by the two-stage statistical procedure following the normality and conservatism tests. More particularly, the source contribution estimation produced by the mixing model has generated high values of GOF and limited levels of uncertainty. The results indicate that reliable estimates of sediment source contribution were obtained by using the fingerprinting technique in this investigation.

The source ascription results presented in **Figure 3** clearly verify that the contribution of the sediments suspended from both the Yangtze mainstream and the Ruxi

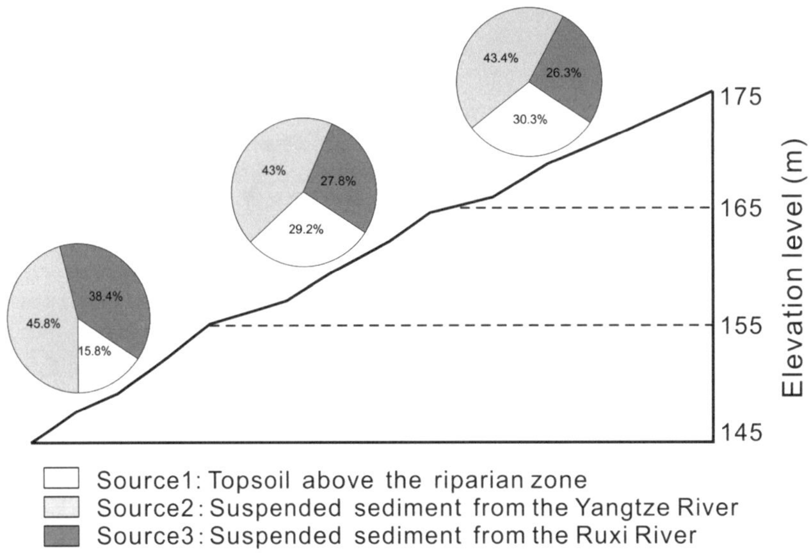

Figure 3.
The median relative contribution of each source type to the deposited sediment collected from the riparian zone with different elevation levels.

tributary channel dominate the sediments deposited in the riparian zone of the reservoir. The sediment contribution originated from these two sources occupied a total proportion of 69.7–84.2%, although there is some evidence of a declining trend with increased elevation. The highest proportional contribution documented for the lowest elevation level of 145–155 m was closely related to the longest water retention time at this level (**Figure 2**), which means that there will be more opportunities for sediment deposition.

In contrast, the proportion of the topsoil source contribution increased from 15.8 to 30.3% with the increasing of elevation. The higher contribution of the topsoil source to the deposits in upper elevation levels (i.e., 155–165 and 165–175 m) could probably be attributed to the proximity of sources to the target sediment sampling sites. During the periods September–October and April–June, which represent the end and the beginning of the wet season in the study area, respectively, sediment originating from the upper lands above the riparian zone caused by water erosion will be preferentially stored in adjacent fields as a result of gentle slope gradient. The relative contribution of this source type to the deposited sediment at the 145–155 m elevation level is therefore much less than those for the upper portions of the riparian zone.

The relatively high sediment contribution from the Ruxi tributary to the 145–155 m deposits might again be explained by the relatively longer submerging period, which last for nearly 9 months (**Figure 2**). Probably more importantly, the sediment mobilized from the upstream catchment and transferred to the Ruxi River during the wet periods September–October and April–June may be predominantly deposited within the 145–155 m level due to the impoundment of the reservoir.

4.2 Environmental implications for sediment source contribution

The information on the relative contribution of potential sediment sources can also be used to evaluate the relative importance of the contributions of sediment-associated nutrients and contaminants [22]. **Figure 4** compares the mean element concentrations for the riverine suspended sediments (sources 2 and 3) and local

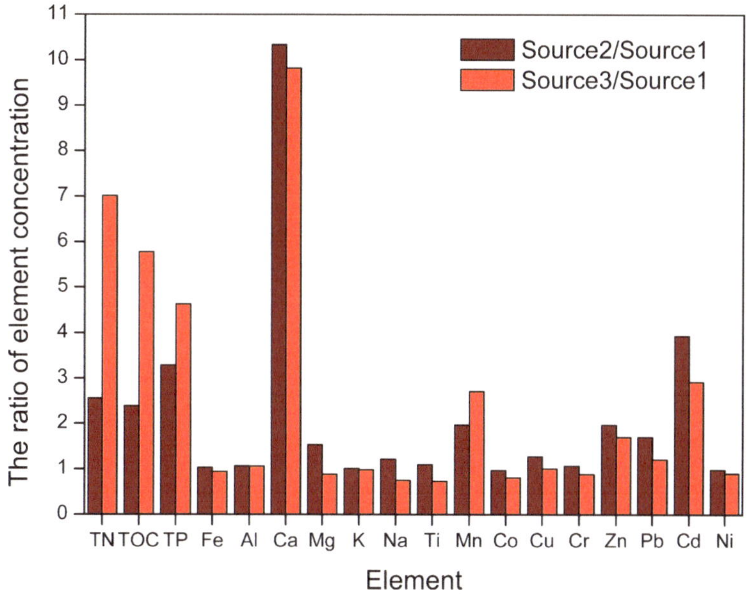

Figure 4.
The ratio of mean concentrations of individual element in suspended sediment collected from the Yangtze River (Source 2) and the Ruxi River (Source 3) to that of the topsoil above the riparian zone (Source 1).

topsoils (Source 1) during the whole sampling period. The presented results show that for all the nutrients (TN, TOC, TP) and some heavy metals (e.g., Mn, Zn, Pb, Cd), the average element concentrations in the suspended fluvial sediments were significantly higher than that one of the local soils. Combining this information with the relative contribution of the suspended fluvial sediments to the riparian deposits, it can be inferred that the sediment-associated nutrients and contaminants transported from both the Yangtze mainstream and the Ruxi tributary represent dominant sources of the contaminants deposited within the riparian zone. Recent studies have documented elevated levels of nutrients [23] and heavy metals [23–25] in the riparian zone of the Three Gorges Reservoir. In this context, the accumulated deposition and storage of fine-grained sediment and associated nutrients and contaminants within the riparian zone of the reservoir may exert negative environmental impacts on aquatic ecosystems and the agricultural use of riparian lands. Moreover, the nutrients and contaminants that stored within the riparian zones may have great potential to be reintroduced into the river system by future bank erosion and/or release to the water column accompanying the annually cyclic soaking.

5. Conclusions

The regular closure in dry season and the water drainage in rainy season of the Three Gorges Dam on the Yangtze River have resulted in significant sedimentation and enrichment of sediment-associated nutrients and contaminants in the reservoir riparian zone. Against this background, the fingerprinting approach was applied to assess the sources of sediment that deposited within the riparian zone of a tributary after the impoundment of the reservoir. Despite significant contrast of sediment contribution from individual sources that was documented for different elevation levels of the riparian zone, the sediments suspended from the Yangtze mainstream was demonstrated to be the primary source of the riparian deposits. From a contamination perspective, however, the sediment input from the upstream tributary also represents an important source of pollution to the riparian environment.

Although it should be recognized that the result reported is limited in temporal and spatial scopes in that only one-year sampling campaign and one single catchment was involved, the findings, combining with the information on previous studies, emphasize the need to take targeted sediment and contaminant management strategies to control the potential environmental problems in the reservoir riparian zones. Further attempts were required to explore the use of sediment fingerprinting approach to assess the sources of sediment-associated nutrients and contaminants in this area.

Acknowledgements

This work was financially supported by the National Natural Science Foundation of China (41430750).

Conflict of interest

The authors have declared that no conflict of interest exists.

Author details

Zhonglin Shi, Dongchun Yan*, Anbang Wen and Yongyan Wang
Institute of Mountain Hazards and Environment, Chinese Academy of Sciences, Chengdu, China

*Address all correspondence to: yandc@imde.ac.cn

References

[1] Bing HJ, Wu YH, Zhou J, Sun HY, Wang XX, Zhu H. Spatial variation of heavy metal contamination in the riparian sediments after two-year flow regulation in the Three Gorges Reservoir, China. Science of the Total Environment. 2019;**649**:1004-1016. DOI: 10.1016/j.scitotenv.2018.08.401

[2] Wang YC, Ao L, Lei B, Zhang S. Assessment of heavy metal contamination from sediment and soil in the riparian zone China's Three Gorges Reservoir. Polish Journal of Environmental Studies. 2015;**24**: 2253-2259. DOI: 10.15244/pjoes/44473

[3] Bao YH, Gao P, He XB. The water-level fluctuation zone of Three Gorges Reservoir – A unique geomorphological unit. Earth-Science Reviews. 2015;**150**:14-24. DOI: 10.1016/j.earscirev.2015.07.005

[4] Yuan XZ, Zhang YW, Liu H, Xiong S, Li B, Deng W. The littoral zone in the Three Gorges Reservoir, China: Challenges and opportunities. Environmental Science and Pollution Research. 2013;**20**:7092-7102. DOI: 10.1007/s11356-012-1404-0

[5] Tang Q, Bao YH, He XB, Fu BJ, Collins AL, Zhang XB. Flow regulation manipulates contemporary seasonal sedimentary dynamics in the reservoir fluctuation zone of the Three Gorges Reservoir. China. Science of the Total Environment. 2016;**548-549**:410-420. DOI: 10.1016/j.scitotenv.2015.12.158

[6] Bao YH, He XB, Wen AB, Gao P, Tang Q, Yan DC, et al. Dynamic changes of soil erosion in a typical disturbance zone of China's Three Gorges Reservoir. Catena. 2018;**169**:128-139. DOI: 10.1016/j.catena.2018.05.032

[7] Yang F, Liu WW, Wang J, Liao L, Wang Y. Riparian vegetation's responses to the new hydrological regimes from the Three Gorges Project: Clues to re vegetation in reservoir water-level-fluctuation zone. Acta Ecologica Sinica. 2012;**32**:89-98. DOI: 10.1016/j.chnaes.2012.02.004

[8] Bing HJ, Zhou J, Wu YH, Wang XX, Sun HY, Li R. Current state, sources, and potential risk of heavy metals in sediments of Three Gorges Reservoir, China. Environmental Pollution. 2016;**214**:485-496. DOI: 10.1016/j.envpol.2016.04.062

[9] MEEP. The Ecological and Environmental Monitoring Bulletin of the Three Gorges Project on Yangtze River. Ministry of Ecology and Environment of the People's Republic of China. 2017:1-51

[10] Walling DE. The evolution of sediment source fingerprinting investigations in fluvial systems. Journal of Soils and Sediments. 2013;**13**:1658-1675. DOI: 10.1007/s11368-013-0767-2

[11] Foster IDL, Lees JA. Tracers in Geomorphology: Theory and applications in tracing fine particulate sediments. In: Foster IDL, editor. Tracers in Geomorphology. Chichester: Wiley; 2000. pp. 3-20

[12] Collins AL, Walling DE, Leeks GJL. Source type ascription of fluvial suspended sediment based on a quantitative composite fingerprinting technique. Catena. 1997;**29**:1-27. DOI: 10.1016/S0341-8162(96)00064-1

[13] Palazón L, Navas A. Variability in source sediment contributions by applying different statistical test for a Pyrenean catchment. Journal of Environmental Management. 2017;**194**:42-53. DOI: 10.1016/j.jenvman.2016.07.058

[14] Collins AL, Pulley S, Foster IDL, Gellis A, Porto P, Horowitz AJ. Sediment

source fingerprinting as an aid to catchment management: A review of the current state of knowledge and a methodological decision-tree for end-users. Journal of Environmental Management. 2017;**194**:86-108. DOI: 10.1016/j.jenvman.2016.09.075

[15] Collins AL, Zhang Y, McChesney D, Walling DE, Haley SM, Smith P. Sediment source tracing in a lowland agricultural catchment in southern England using a modified procedure combining statistical analysis and numerical modelling. Science of the Total Environment. 2012;**414**:301-317. DOI: 10.1016/j.scitotenv.2011.10.062

[16] Smith HG, Blake WH. Sediment fingerprinting in agricultural catchments: A critical re-examination of source discrimination and data corrections. Geomorphology. 2014;**204**:177-191. DOI: 10.1016/j.geomorph.2013.08.003

[17] Koiter AJ, Owens PN, Petticrew EL, Lobb DA. Assessment of particle size and organic matter correction factors in sediment source fingerprinting investigations: An example of two contrasting watershed on Canada. Geoderma. 2018;**325**:195-207. DOI: 10.1016/j.geoderma.2018.02.044

[18] Wilkinson S, Hancock G, Bartley R, Hawdon A, Keen R. Using sediment tracing to assess processes and spatial patterns of erosion in grazed rangelands, Burdekin River basin, Australia. Agriculture Ecosystems & Environment. 2013;**180**:90-102. DOI: 10.1016/j.agee.2012.02.002

[19] Nosrati K, Collins AL, Madankan M. Fingerprinting sub-basin spatial sediment sources using different multivariate statistical techniques and the modified MixSAR model. Catena. 2018;**164**:32-43. DOI: 10.1016/j.catena.2018.01.003

[20] Laceby JP, Olley J. An examination of geochemical modelling approaches to tracing sediment sources incorporating distribution mixing and elemental correlations. Hydrological Processes. 2015;**29**:1669-1685. DOI: 10.1002/hyp.10287

[21] Motha JA, Wallbrink PJ, Hairsine PB, Grayson PB. Determining the sources of suspended sediment in a forested catchment in southeastern Australia. Water Resources Research. 2003;**39**:1056-1069. DOI: 10.1029/2001wr000794

[22] Walling DE, Collins AL, Stroud RW. Tracing suspended sediment and particulate phosphorus sources in catchments. Journal of Hydrology. 2008;**350**:274-289. DOI: 10.1016/j.jhydro.2007.10.047

[23] Shi ZL, Wang YY, Wen AB, Yan DC. Chen JC. Temporal-spatial variations of sediment-associated nutrients and contaminants in the Ruxi tributary of the Three Gorges Reservoir, China. Journal of Mountain Science. 2018;**15**:319-326. DOI: 10.1007/s11629-017-4486-9

[24] Ye C, Li SY, Zhang YL, Zhang QF. Assessing soil heavy metal pollution in the water-level-fluctuation zone of the Three Gorges Reservoir, China. Materials. 2011;**191**:366-372. DOI: 10.1016/j.jhazmat.2011.04.090

[25] Tang Q, Bao YH, He XB, Zhou HD, Cao ZJ, Gao P, et al. Sedimentation and associated trace metal enrichment in the riparian zone of the Three Gorges Reservoir, China. Science of the Total Environment. 2014;**479-480**:258-266. DOI: 10.1016/j.scitotenv.2014.01.122

Chapter 5

A Long-Term Prediction of Beach Changes around River Delta using Contour-Line-Change Model

Takaaki Uda, Shiho Miyahara, Toshiro San-nami and Masumi Serizawa

Abstract

The long-term beach changes along the Kaike coast (Japan) have been investigated. Being a major source of sand to this coast, the Hino River has supplied a large amount of sand during the extensive mining of iron sand in the past, resulting in the shoreline advance. However, the stoppage of the sand mining has caused a marked decrease in the sand supply, resulting in a rapid shoreline recession around the river delta. The beach changes triggered by these human activities have been investigated using an old geographical map and aerial photographs taken between 1947 and 2005. Then, the beach changes have been reproduced using the contour-line-change model considering the change in grain size of the beach sediments. Bathymetric data have been analyzed in order to evaluate the longshore sand transport and the fluvial sand supply from the Hino River. The measured and predicted three-dimensional beach changes were in good agreement and the effectiveness of the contour-line-change model for predicting long-term beach changes was confirmed by this case history.

Keywords: beach changes, contour-line-change model, river delta, longshore sand transport, Kaike coast

1. Introduction

The Yumigahama Peninsula located in the western part of Tottori Prefecture, Japan, has developed as a tombolo owing to the wave-sheltering effect of the Shimane Peninsula with a rich sand supply from the Hino River. The 18-km-long coastline between the Yodoe fishing port and Sakai Port located at both the ends of the peninsula is called the Kaike coast (**Figure 1**). In the Hino River basin, a large-scale mining of iron sand had been carried out in the nineteenth and early twentieth centuries. A huge amount of sand was abandoned in the river basin and such a sand was transported toward the coast through the Hino River. With the cessation of the iron sand mining, however, the sand discharge from the Hino River markedly decreased, causing a strong erosion around the river mouth. The erosion on the Kaike coast was severe from the early 1930s to immediately after World War II. At first, groins were constructed as a measure against erosion, and the construction of detached breakwaters began in 1971. Behind 12 detached breakwaters constructed by 1982, cuspate forelands were well developed [1–3]. Although these detached

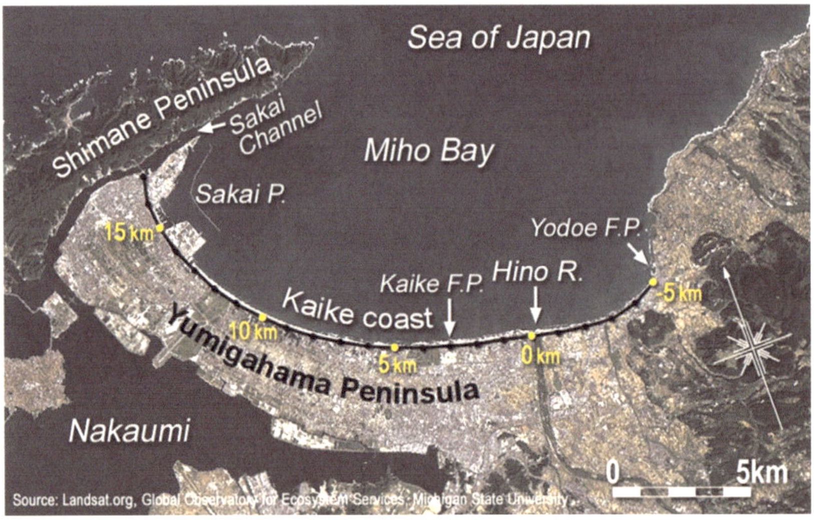

Figure 1.
Location of Yumigahama Peninsula in Tottori Prefecture, Japan.

breakwaters and cuspate forelands have been stably maintained, fine-grained sand was locally deposited, natural sandy beaches rapidly disappeared, and the coastline was covered by artificial structures in the entire Yumigahama Peninsula [4].

In predicting the beach changes triggered by the imbalance in longshore sand transport on a coast such as the Kaike coast, a long-term prediction in an extensive area is required [5]. In such a case, the time scale changes from years to decades and the calculation domain reaches even up to 10–100 km. The N-line model can be applicable to the prediction of such beach changes [6–12]. We have developed the contour-line-change model considering the changes in grain size of the seabed materials as a type of N-line model to predict long-term beach changes including the prediction of changes in grain size [13, 14]. The model has applied to the long-term prediction of beach changes around a river mouth delta and those along an arc-shaped shoreline in previous papers.

Uda et al. [15] have first shown a predictive model of the three-dimensional development and deformation of a river mouth delta. This model enabled the prediction of not only the shoreline change but also the three-dimensional, long-term topographic changes around a river mouth, and offshore sand transport can be taken into account when the sea bottom slope exceeds a critical value given by the angle of repose of sand. Furuike et al. [16] have applied the model to the Enshu-nada coast. The shoreline changes between the Magome river mouth and Imagire-guchi jetty have been investigated where beach changes triggered by the decrease in fluvial sand supply from the Tenryu River occur. It was observed that the shoreline retreated downcoast of the Magome river mouth, and the shoreline advanced in parallel further downcoast, while maintaining the curvature in the shoreline configuration. Miyahara et al. [17, 18] have investigated the long-term evolution of the Tenryu River delta associated with sand bypassing at several dams. When the sediment yield from the river was artificially increased, the supplied sediment was mainly deposited around the river mouth, but it took a longer time for a sandy beach far from the river mouth to recover. Given the annual discharge of sediment with three grain sizes, the recovery of the delta topography and the effect of nourishment on the nearby coast have been predicted. San-nami et al. [19] have analyzed the beach changes of Kujukuri Beach with a 60 km length located in Boso Peninsula. On south Kujukuri Beach, severe beach erosion has occurred since the

1970s, and the erosion area has been expanding. The exhaustion of sand supply by northward longshore sand transport from the sea cliffs, formation of a wave-shelter zone at Katakai fishing port breakwater, and ground subsidence caused by the pumping-up of underground water were the major causes of the shoreline recession. The arc-shaped shoreline of Kujukuri Beach was reproduced given these conditions. Furthermore, San-nami et al. [20] have investigated the long-term topographic changes since 1968 along the entire Shizuoka and Shimizu coasts including a 17 km stretch extending between the Abe River and the Mihono-matsubara sand spit. The beach erosion of these coasts was triggered by the decrease in sediment supply from the Abe River due to excessive riverbed mining until 1967. After 1982/1983, the natural sand supply from the river increased and accretion occurred on these coasts. Measured topographic changes were reproduced using the model. Not only the movement of the sand body but also the shoreline and bathymetric changes were numerically reproduced.

Upon these previous studies, the long-term topographic changes of the Kaike coast in a 70-year period have been analyzed in this study, including the period when extensive mining of iron sand has been carried out. Then, given these conditions, the process of the reduction in the size of the Hino River delta over 70 years was reproduced using the contour-line-change model considering the change in grain size of the seabed material, and the applicability of the model was validated.

2. Long-term shoreline changes along Yumigahama Peninsula and longshore sand transport

2.1 Shoreline changes

The shoreline configurations in the Yumigahama Peninsula were determined from an old geographical map produced in 1899 and the aerial photographs taken in 1947, 1967, and 1973. The shoreline changes in 74 years were investigated between 1899, when rich sediment was supplied from the Hino River during the period of iron sand mining, and 1973, when the construction of detached breakwaters began. In the analysis of the shoreline changes, the correction owing to the changes in tide level was made using the tide level when the aerial photograph was taken and the mean foreshore slope of 1/15. The origin of the coordinate (x = 0 km) was set at the left bank of the Hino River and the alongshore distance x was taken westward along the coastline (**Figure 1**).

Figure 2 shows the shoreline changes between 1899 and 1947. The shoreline markedly retreated in an area between x = 1 and $-x$ = 2 km around the Hino River mouth with the decline of mining of iron sand, and the shoreline receded by a maximum of 250 m at the river mouth. On the other hand, in the areas westwards and eastwards of x = 3 km, the shoreline advanced of 100 m in parallel at a rate of 2 m/year, except that in the vicinity of Yodoe fishing port. Similarly, **Figure 3** shows the shoreline changes with reference to shoreline in 1947 until 1967 and 1973, after that two detached breakwaters were constructed. Since a seawall and groins have been constructed in an area between x = 0 and x = 1.5 km westwards of the Hino River until 1967, the shoreline recession was prevented. However, excluding this area, the shoreline retreated westwards of the protected area. The eroded sand was transported westwards of x = 5 km, resulting in the shoreline advance. Until 1973, the area where the shoreline markedly receded expanded westwards, and the area where the shoreline advanced until 1967 was eroded. Here, the shoreline advance at two locations near x = 1 km was due to the formation of cuspate forelands behind the detached breakwaters constructed by 1973.

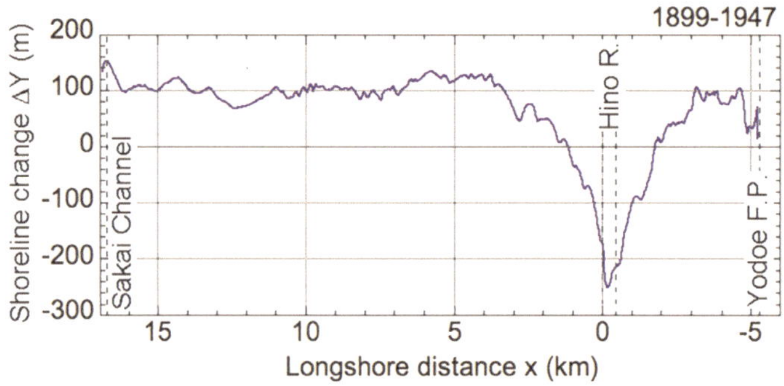

Figure 2.
Shoreline changes in entire Yumigahama Peninsula (reference year: 1899).

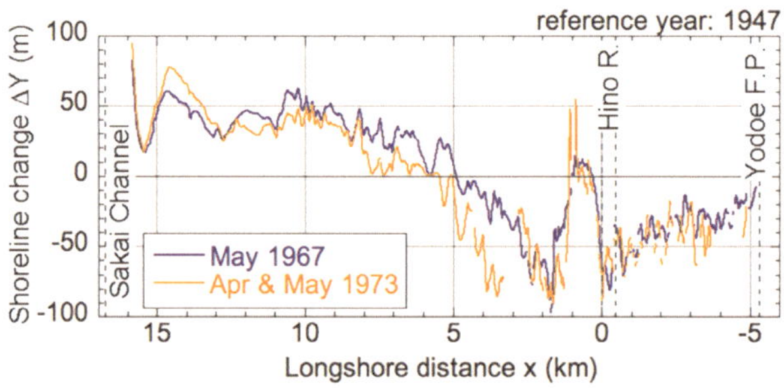

Figure 3.
Shoreline changes until 1967 and 1973 in entire Yumigahama Peninsula (reference year: 1947).

2.2 Analysis of bathymetric survey data

Using the bathymetric survey data (200 m intervals), the bathymetry in 2000 and the bathymetric changes between 1980 and 2000 have been represented (**Figures 4** and **5**). The eroded and accreted sand volumetric changes relative to those occurring in 1980 are also shown together with the locations of the detached breakwaters (DBs; **Figure 5**). The contours shallower than −8 m markedly protruded near the river mouth delta, and the DBs were constructed immediately westwards of this river delta. Twelve DBs were constructed from 1971 to 1982. With the construction of DBs, sand was deposited behind these breakwaters, whereas erosion occurred in the offshore zone between −4 and − 7 m, and down the coast of the accretion area. These beach changes were considered to be due to the blocking of westward alongshore sand transport by the DBs.

2.3 Longitudinal profiles and depth change in median diameter

In July 2002, the sampling of the seabed material was carried out in the depth zone between +3 and − 12 m along transect x = 1.3 km at 1-m-depth intervals. **Figure 6** shows the longitudinal profile and depth distribution of median diameter (d_{50}) along this transect. Coarse sand of d_{50} ranging from 0.3 to 0.7 mm is deposited on the foreshore, and d_{50} decreases with increasing depth, and well-sorted fine sand with a grain size of 0.13 mm is deposited in the depth zone deeper than −8 m. Corresponding to this grain size distribution, the seabed slope is as steep as 1/6 on the foreshore, 1/30 between −2 and − 8 m, and a very gentle slope of 1/100 in the depth zone deeper than −8 m. From these data, it is found that the depth of closure (h_c) is approximately 8 m on this coast, and the seabed near h_c is covered by fine

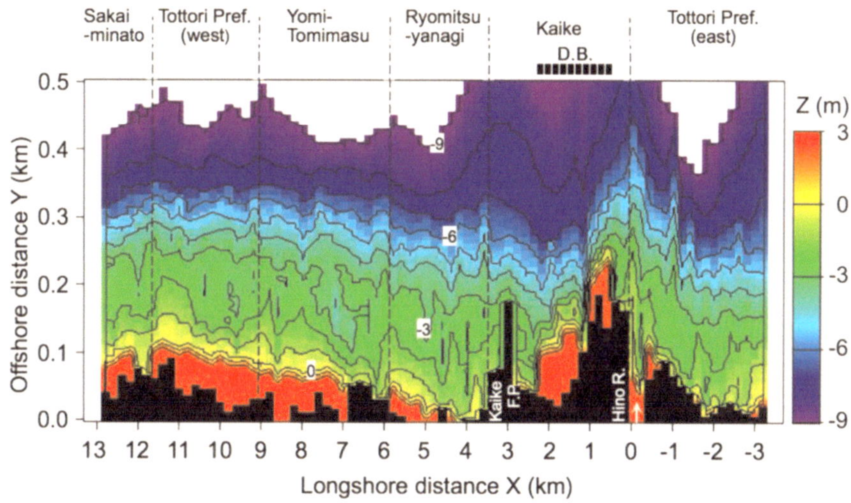

Figure 4.
Bathymetry in 1980 using expanded coordinates.

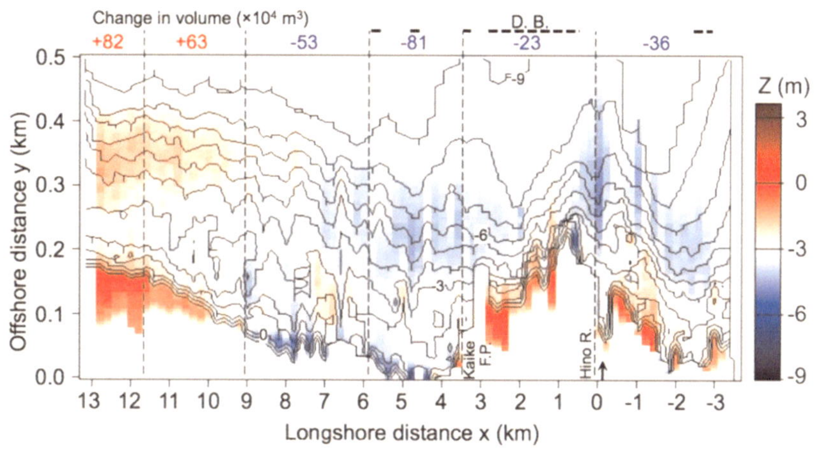

Figure 5.
Bathymetry in 2000 using expanded coordinates and bathymetric changes with reference to that in 1980.

sand. The depth distribution of d_{50} can be separated into three subregions, as shown in **Figure 6**, with longitudinal slopes of 1/6, 1/30, and 1/100, i.e., fluvial sediment supplied from the Hino River is mainly distributed in the depth zone corresponding to the grain size of the material, and coarser material is deposited near the shoreline, and the component finer than the grain size of 0.13 mm is deposited in the zone deeper than −8 m. In addition, fine sand of d_{50} = 0.3 mm is deposited in the intermediate depth zone between −2 and − 8 m. Although westward longshore sand transport prevails on the coast, sand is transported along the depth zones separated in two layers associated with the grain size sorting (**Figure 6**). Thus, the changes in contour lines in the depth zones I and II between z = +3 and − 2 m and between z = −2 and − 8 m could be separately calculated, where the average grain sizes are 0.5 and 0.3 mm, respectively.

Figure 7 shows the longshore distribution of changes in sand volume in the depth zones I and II between 1980 and 2009. In the depth zone I, the sand volume was almost constant east of x = 5 km except for the local deposition of sand behind

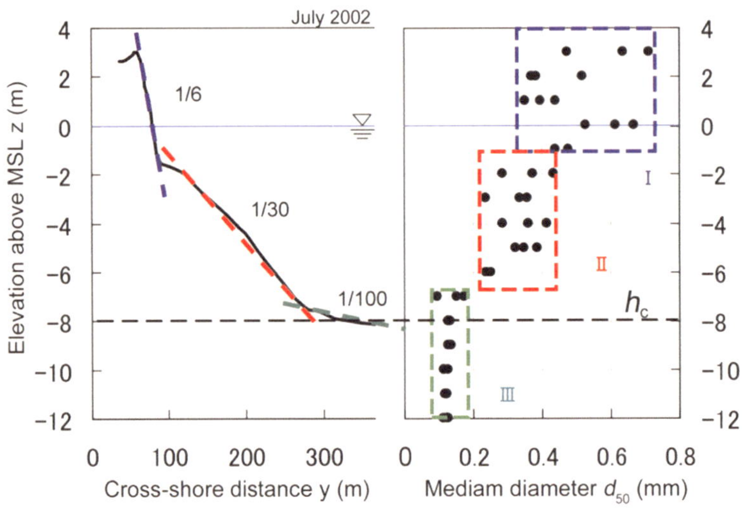

Figure 6.
Longitudinal profile along transect x = 1.3 km and depth distribution of d_{50} measured in July 2002.

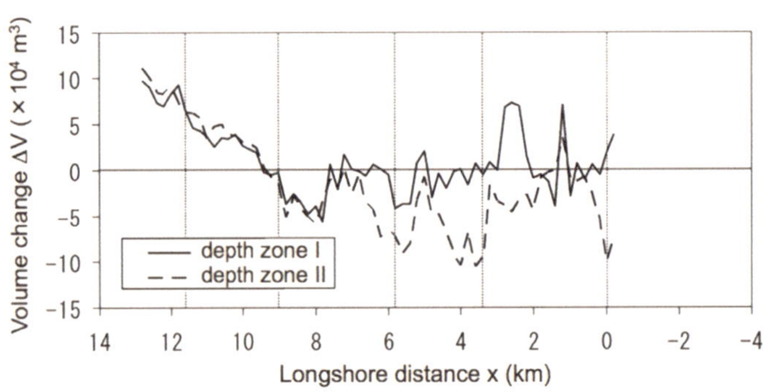

Figure 7.
Longshore distribution of changes in sand volume in depth zones I and II between 1980 and 2009.

detached breakwaters constructed between $x = 0$ and 4 km, and the accreted sand volume was 1.34×10^6 m^3 in contrast to the eroded sand volume of 5.8×10^5 m^3. The sand volume in the depth zone II decreased between $x = 2$ and 9 km, whereas it increased west of $x = 9$ km; the eroded volume was 2.01×10^6 m^3 greater than the accreted volume of 1.06×10^6 m^3. However, the eroded and accreted sand volumes in the entire depth zones I and II were 2.59×10^6 and 2.40×10^6 m^3, respectively, and the accreted volume accounts for 93% of the eroded volume.

2.4 Longshore sand transport

The west and east ends of the study area are the solid boundaries of the reclaimed land and Yodoe fishing port; thus, the distribution of longshore sand transport can be calculated by integrating the shoreline change from these boundaries to the Hino River mouth, and multiplying by the characteristic height of beach changes, which is defined by the correlation coefficient between the change in the cross-sectional area of the beach and the shoreline change, and then divided by the elapsed time. Since the relationship of $h = (1.0{\sim}1.3)h_c$ holds between the characteristic height of beach changes (h) and h_c [4], and h_c on this coast is approximately equal to 8 m, h becomes 10 m.

DOI: http://dx.doi.org/10.5772/intechopen.85207

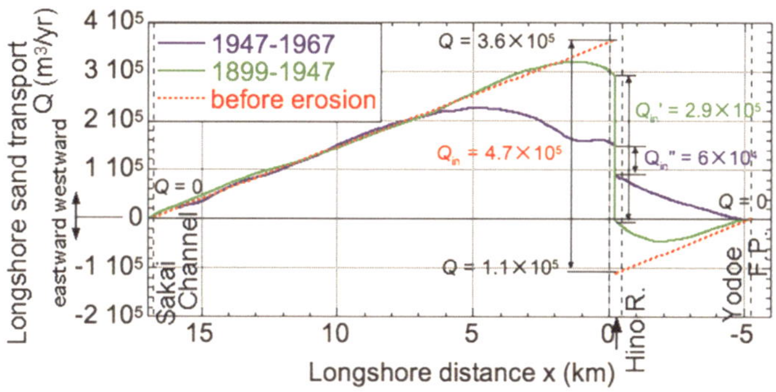

Figure 8.
Distribution of longshore sand transport between 1899 and 1947 and between 1947 and 1967.

Figure 8 shows the distribution of longshore sand transport between 1899 and 1947, and between 1947 and 1967 with a positive sign for westward longshore sand transport. Although westward and eastward longshore sand transports took place on the west and east sides of the river mouth between 1899 and 1947, respectively, the distribution of longshore sand transport changes linearly with the same slope west of $x = 3$ km and east of $x = -3$ km, and they smoothly converge to the distribution between 1947 and 1967 west of $x = 7$ km. Westward and eastward longshore sand transports take maximum rates of 3.2×10^5 and 4.6×10^4 m^3/year at $x = 1$ and $-$ 1.8 km, respectively. On both sides of these points, the longshore sand transport is linearly distributed, so that it is found that the shoreline has advanced in parallel over time except for the erosion area close to the river delta. The longshore sand transport has a difference of 2.9×10^5 m^3/year between both sides of the river mouth, implying that the same amount of sand was supplied from the river during the period. In contrast, longshore sand transport became westward in the entire area between 1947 and 1967 after the fluvial sand supply was markedly reduced owing to the stoppage of the mining of iron sand with a maximum rate of 2.3×10^5 m^3/year at $x = 5$ km. The location where the longshore sand transport takes a peak value shifted westward compared with that between 1899 and 1967, implying that the erosion area expanded westward. Furthermore, since longshore sand transport on the west and east sides of the river mouth become 1.5×10^5 and 9.0×10^4 m^3/year, respectively, sediment inflow from the river results in 6.0×10^4 m^3/year. In addition, in **Figure 8**, the longshore sand transport linearly increased from the west end to a location of $x = 7$ km in both periods between 1899 and 1947 and between 1947 and 1967, and the effect of the decrease in the longshore sand transport did not reach far from the river mouth, resulting in the parallel advance of the shoreline. Therefore, when this straight line is extrapolated toward the river mouth, a straight line indicated by "before erosion" in **Figure 8** is obtained, and the longshore sand transport at the river mouth became 3.6×10^5 m^3/year, which is assumed to be the westward longshore sand transport before the initiation of erosion.

In the area east of the river mouth, eastward longshore sand transport could have prevailed with the development of the river delta, and the longshore sand transport is assumed to have a linear distribution with the same longshore inclination as that west of the river mouth to satisfy the conditions that the shoreline around the river mouth is continuous and the river delta parallelly advanced in the offshore direction. By extrapolating a straight line with the same inclination as that west of the river mouth from Yodoe fishing port, we obtained a longshore sand transport of 1.1×10^5 m^3/year on the east side of the river mouth. Finally, the entire sediment supply from the river in the period when iron sand was mined is assumed

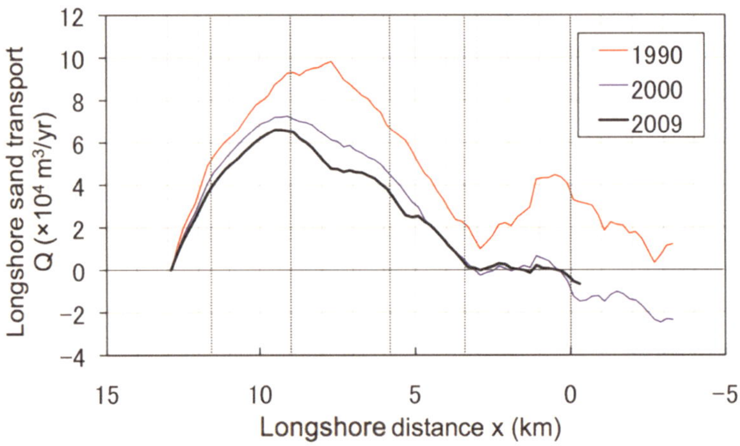

Figure 9.
Distribution of longshore sand transport calculated from beach changes between 1980 and 2009.

to be 4.7×10^5 m^3/year as the sum of longshore sand transport on both sides of the river mouth. In addition, since the longshore inclination of longshore sand transport in **Figure 8** gives 21 m^3/m/year as the rate of sand accretion, the rate of the advance of the river delta becomes 2.1 m/year, when dividing this rate by the characteristic height of beach changes of h = 10 m.

Similarly, the distribution of longshore sand transport after 1980 was evaluated by integrating the topographic changes between 1980 and 2009 in the east from the west end, and dividing by the elapsed time. **Figure 9** shows the results. The longshore sand transport took a maximum of 1.0×10^5 m^3/year at x = 7.5 km in 1990 with a secondary peak at x = 0 km. After 2000, it has decreased in the eastern part, and it took a maximum of 6.5×10^4 m^3/year in 2009.

3. Change in apex angle of Hino River delta and predominant wave direction

Figure 10 shows the geographical map in 1899 and aerial photograph in 1947. The apex angle of the river delta (angle AOB) in 1899 was 170°. Assuming that the wave crest line of the predominant wave incident to the river delta is given by a straight line of A′OB′ in **Figure 10(a)**, the breaker angles between the shoreline on both sides of the river delta and the wave crest line are given by AOA′ and BOB′. These breaker angles can be evaluated as follows. First, the sum of these breaker angles becomes 10° by subtracting the apex angle of the river delta of 170° from 180°. In the distribution of the longshore sand transport shown in **Figure 8**, the longshore sand transport on the west and east sides of the Hino River mouth are 3.6×10^5 and 1.1×10^5 m^3/year, respectively, resulting in the ratio of longshore sand transport on both sides of 3:1, because the longshore sand transport is proportional to the breaker angle. Taking these conditions into account, the breaker angles on the left and right banks of the river mouth decrease to 7.5° and 2.5°, respectively, when separating 10° by the ratio of 3:1. Finally, the predominant wave direction normal to A′B′ is assumed to be N13°E.

In 1947, the shoreline of the river delta receded because of the decrease in sand supply from the river, as shown in **Figure 10(b)**, and the apex angle increased to up to 175°, and the breaker angles on the left and right banks of the river became 5° and 0°, respectively, taking the predominant wave direction of N13°E calculated as

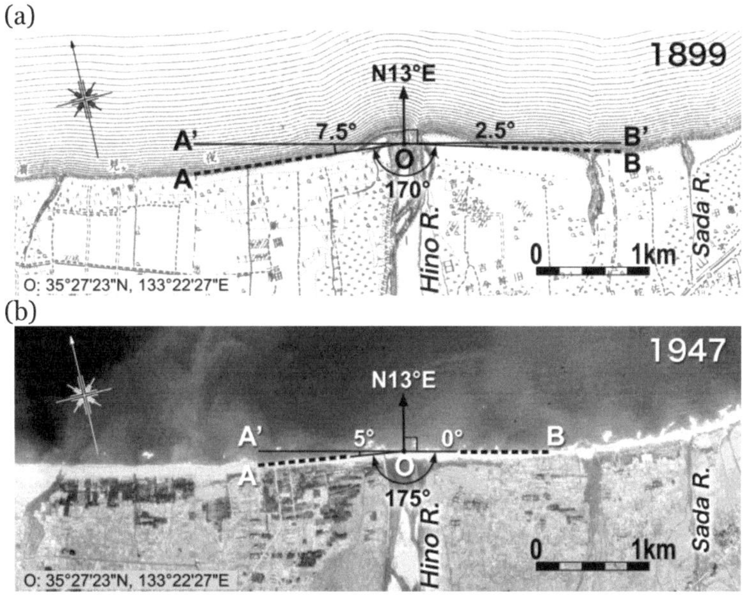

Figure 10.
Change in apex angle of Hino River delta and predominant wave direction. (a) 1899. (b) 1947.

mentioned above. The decrease in the breaker angle west of the river mouth corresponds to the decrease in the westward longshore sand transport.

4. Contour-line-change model considering change in grain size

A numerical model described in [13, 14] was used. Let the x- and z-axes be the longshore distance and seabed elevation relative to the still water level, respectively, and Y (x, z, t) is the offshore distance to a specific contour line to be solved. To consider the sorting of sand of different grain sizes by cross-shore sand transport, the depth distribution of cross-shore sand transport considering the grain size effect in mixing was considered. The sorting of grain size populations was modeled by introducing the equilibrium slope angle $\beta_c{}^{(k)}$, which corresponds to each grain size population k. In this case, a grain size population was assumed to have a single equilibrium beach slope with a characteristic grain size $d^{(k)}$. By assuming that the mobility of sand of each grain size population by cross-shore movement is the same as that of longshore sand transport, the coefficient of the sediment transport rate according to the grain size $d^{(k)}$, which was given in [21, 22], was introduced. Furthermore, assuming that the ratio of the exposed area of each grain size population to the entire sea bottom area is equal to the content of each size population in the exchange layer $\mu^{(k)}(k = 1, 2, \cdots, N)$, the cross-shore sand transport of each grain size population $q_z{}^{(k)}$ is derived.

Cross-shore sand transport:

$$q_z^{(k)} = \mu^{(k)} \cdot \varepsilon_z(z) \cdot \gamma \cdot K_1^{(k)} \cdot \left(EC_g\right)_b \cos^2\alpha_b \sin\overline{\beta} \cdot \left(\cot\beta / \cot\beta_c^{(k)} - 1\right); \quad k = 1, 2, \cdots, N \tag{1}$$

$$K_1^{(k)} = \frac{A}{\sqrt{d^{(k)}}}; k = 1, 2, \cdots, N \tag{2}$$

$$\cot\beta = -\partial Y/\partial z \tag{3}$$

$$\varepsilon_x(z) = \begin{cases} (2/h_c^3)(h_c/2 - z)(z + h_c)^2, & -h_c \le z \le h_R \\ 0, z \le -h_c, z \ge h_R \end{cases} \tag{4}$$

Here, $q_z^{(k)}(k = 1, 2, \cdots, N)$ is the cross-shore sand transport per unit length in the longshore direction for each grain size population, $\mu^{(k)}$ is the content of each grain size population (k) in the exchange layer of sand, $(EC_g)_b$ is the wave energy flux at the breaking point, $K_1{}^{(k)}$ is the coefficient of longshore sand transport, $\varepsilon_z(z)$ is assumed to be equivalent to the depth distribution of the longshore sand transport $\varepsilon_x(z)$ given by Uda and Kawano [23], and $d^{(k)}$ is a typical grain size of the grain size population. A is a coefficient that depends on the physical conditions of the beach. $d^{(k)}$ in Eq. (2) has a unit of mm. γ is the ratio of the coefficient of cross-shore sand transport to that of longshore sand transport ($\gamma = K_z{}^{(k)}/K_1{}^{(k)}$) and expresses the mobility of cross-shore sand transport relative to that of longshore sand transport, α_b is the angle between the wave crest line at the breaking point and each contour line, and β is the beach slope angle at each contour line. $\overline{\beta}$ is the average beach slope angle between the berm height h_R and the depth of closure h_c, and $\beta_c{}^{(k)}$ is the equilibrium beach slope angle. When the beach slope becomes steeper than the angle of repose of sand, sand is transported offshore by gravity. By this procedure, we can calculate the formation of a scarp in a zone larger than h_R and the sinking of sand in a zone larger than h_c.

Longshore sand transport:

$$q_x{}^{(k)} = \mu^{(k)} \cdot \varepsilon_x(z) \cdot K_1^{(k)} \cdot (EC_g)_b \cdot \left(\cos\alpha_b \sin\alpha_b - \xi \frac{1}{\tan\beta_b} \cdot \cos\alpha_b \cdot \frac{\partial H_b}{\partial x} \right); \tag{5}$$

$$k = 1, 2, \cdots, N$$

Here, $q_x^{(k)}(k = 1, 2, \cdots, N)$ is the longshore sand transport per unit depth for each grain size population, □$\varepsilon_x(z)$ is the depth distribution of longshore sand transport, and ξ is the constant given by $K_2{}^{(k)}/K_1{}^{(k)}$, which depends on the physical conditions of the beach, where $K_2{}^{(k)}$ is a function of $K_1{}^{(k)}$ and is equivalent to the coefficient of Ozasa and Brampton [24]. $\tan\beta_b$ is the beach slope in the surf zone and H_b is the breaker height.

Mass conservation for each grain size:

$$\frac{\partial y^{(k)}}{\partial t} = -\frac{\partial q_x^{(k)}}{\partial x} - \frac{\partial q_z^{(k)}}{\partial z} + S^{(k)}; \tag{6}$$

$$k = 1, 2, \cdots, N$$

Here, $S^{(k)}(k = 1, 2, \cdots, N)$ is the additional term for calculating the sand source for each grain size population, by which sediment supply from a river or the sand supply by beach nourishment is given. The total contour-line change at a certain position is determined by the summation of the contour-line changes of all grain size populations at that position.

$$\frac{\partial Y}{\partial t} = \sum_{k=1}^{N} \frac{\partial y^{(k)}}{\partial t} \tag{7}$$

Change in content of each grain size population:

$$\frac{\partial \mu^{(k)}}{\partial t} = \frac{1}{B}\left\{\frac{\partial y^{(k)}}{\partial t} - \frac{\partial Y}{\partial t}\cdot \mu^{(k)}\right\},\ \frac{\partial Y}{\partial t} \geq 0; \quad k = 1, 2, \cdots, N \tag{8}$$

The content of each grain size population in the new exchange layer formed during erosion is expressed as:

$$\frac{\partial \mu^{(k)}}{\partial t} = \frac{1}{B}\left\{\frac{\partial y^{(k)}}{\partial t} - \frac{\partial Y}{\partial t}\cdot \mu_B^{(k)}\right\},\ \frac{\partial Y}{\partial t} \leq 0; \quad k = 1, 2, \cdots, N \tag{9}$$

where $\mu_B^{(k)}$ is the content of each grain size population on the sandy beach landward of the initial exchange layer. The width B of the exchange layer is determined with reference to the mixing depth reported by Kraus [25]. The abovementioned equations were solved simultaneously. In the numerical simulation, the calculation points of the contour-line position and sand transport rate are set in staggered meshes with a difference of 1/2 mesh; the cross-shore and longshore sand transport are calculated using Eqs. (1) and (5), respectively. In this case, the upcoast value depending on the direction of sand transport is employed as $\mu^{(k)}$ in Eqs. (1) and (5). The bathymetric changes can be calculated using Eqs. (6) and (7). $\mu^{(k)}$ is calculated using Eqs. (8) and (9), and the calculation point of $\mu^{(k)}$ is set at the same point for calculating the contour-line position.

5. Reproduction of reduction in size of Hino River delta

5.1 Calculation conditions

The numerical simulation of topographic changes relative to the initial topography in 1899 was carried out until 1967, when a large amount of sand was supplied from the river as a result of the extensive mining of iron sand, and then sand supply (Q_{in}) markedly decreased after the stoppage of the mining. The calculation period was separated into four periods corresponding to the sand discharge and the structural conditions of the coast. In the first period between 1899 and 1920, Q_{in} of 4.7×10^5 m^3/year was supplied from the river, resulting in the development of the river delta. In this period, the shoreline in the entire area advanced at a rate of 2 m/year, and the entire shoreline advance reached 42 m by 1920.

In the second period between 1920 and 1947, Q_{in} began to decrease from 4.7×10^5 m^3/year before 1920 to 1.5×10^5 m^3/year after 1920. Q_{in} of 1.5×10^5 m^3/year was calculated by subtracting the sand volume in the first period (4.7×10^5 m^3/year $\times$ 21 year) from that between 1899 and 1947 (2.9×10^5 m^3/year $\times$ 48 year), which was determined from the shoreline changes between 1899 and 1947, and divided by the elapsed time of 27 years. In the third period between 1947 and 1962, Q_{in} decreased until 6×10^4 m^3/year owing to the stoppage of mining of iron sand, which was calculated from the measured shoreline changes between 1947 and 1967. In the fourth period between 1962 and 1967, the seawall and groins were constructed west of the Hino River mouth, although Q_{in} remained constant at 6×10^4 m^3/year.

By setting the coastline lengths in the areas west and east of the river mouth to be L_1 and L_2, we found that the wave directions $-\theta_1$ and θ_2 immediately west and east of the river mouth, respectively, satisfy a relationship of $-\theta_1/\theta_2 = L_1/L_2 = 3$. In

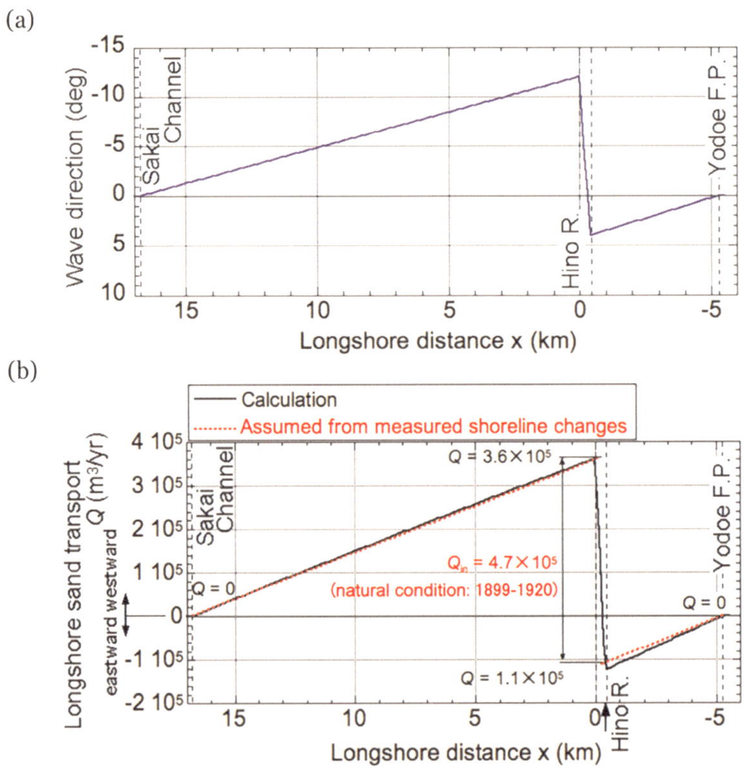

Figure 11.
Wave direction and distribution of longshore sand transport used for calculation. (a) Wave direction. (b) Longshore sand transport.

addition, $-\theta_1$ and θ_2 are approximately given by 7.5° and 2.5°, respectively, from the geographical map in 1899, as shown in **Figure 10**. Thus, after the trial-and-error calculation of the shoreline changes, while satisfying the relation of $\theta_1/\theta_2 = 3$, we obtained $-\theta_1 = 12°$ and $\theta_2 = 4°$ for the best fit results. Finally, a linear distribution shown in **Figure 11(a)** was assumed with angles of −12° and 4° immediately west and east of the Hino River mouth, respectively. **Figure 11(b)** shows the distribution of longshore sand transport at the initial stage. For the wave condition, the breaker height was determined as $H_b = 1.1$ m on the basis of the energy mean significant wave height measured between 1995 and 2008 at Hiezu wave observatory.

The initial seabed slope was assumed to be 1/6 between $z = +3$ and − 2 m, and 1/30 between $z = -2$ and − 8 m. The number of grain sizes (N) was set to 2 with characteristic grain sizes of $d^{(1)} = 0.3$ mm for fine sand and $d^{(2)} = 0.5$ mm for coarse sand. The initial contents μ_1 and μ_2 for fine and coarse sand were, respectively, assumed to be 0.0 and 1.0 in the cell between $z = +3$ and − 1 m, 0.5 and 0.5 in the cell at $z = -2$ m, and 1.0 and 0.0 in the cell between $z = -3$ and − 8 m. The equilibrium slopes $\tan\beta_c^{(1)}$ and $\tan\beta_c^{(2)}$ of fine and coarse sand were, respectively, assumed to be 1/30 and 1/6, on the basis of the measured longitudinal profile. As for the sand back pass on the coast, sand was extracted from the foreshore with an elevation between 0 and + 3 m at $x = 13$ km, and the same amount of sand was supplied from the foreshore at $x = 8$ km. **Table 1** summarizes the other calculation conditions.

5.2 Results of numerical simulation

Figure 12 shows the bathymetric changes from the first to the fourth periods together with an additional expression of bathymetric changes relative to the

Calculation method	Contour-line-change model considering grain size changes
Calculation domain	Sakai channel to Yodoe fishing port
Q_{in} in each calculation period	Reproduction calculation between 1899 and 1967 1 (1899–1920): 4.7×10^5 m^3/year 2 (1920–1947): 1.5×10^5 m^3/year 3 (1947–1962): 6×10^4 m^3/year 4 (1962–1967): 6×10^4 m^3/year + seawall
Initial bathymetry	Straight parallel contours between z = +3 m and − 2 m with 1/6 slope, between z = −2 m and − 8 m, 1/30 slope
Grain size	• $N = 2$ • For characteristic grain sizes, fine sand of $d^{(1)}$ = 0.3 mm coarse sand of $d^{(2)}$ = 0.5 mm • Initial contents z = 3 m to −1 m: μ_1 = 0.0 (fine sand) and μ_2 = 1.0 (coarse sand) z = −2 m: μ_1 = 0.5 (fine sand) and μ_2 = 0.5 (coarse sand) z = −3 m to −8 m: μ_1 = 1.0 (fine sand) and μ_2 = 0.0 (coarse sand)
Equilibrium slope	$\tan\beta_c^{(1)}$ = 1/30 and $\tan\beta_c^{(2)}$ = 1/6
Width of exchange layer	B = 6 m (thickness of exchange layer = 1 m)
Incident wave condition	• Breaker height: H_b = 1.1 m, energy-mean significant wave height measured between 1995 and 2008 at Hiezu observatory • Initial breaker angle: linear distribution between 0° (Sakai channel) and − 12° (Hino River mouth), and between 4° (Hino River mouth) and 0° (Yodoe fishing port)
Tide condition	MSL
Depth of closure and berm height	Depth of closure h_c = 8 m Berm height h_R = 3 m
Coefficient of sand transport	Coefficient of longshore sand transport A = 0.12 for initial longshore sand transport to be $Q = 3.6 \times 10^5$ m^3/year Ratio of cross-shore and longshore sand transports γ = 0.1 Coefficient of Ozasa and Brampton's [18] term ξ = 3.24
Depth distribution of longshore and cross-shore sand transport	Uniform
Critical slope of falling sand	On land: 1/2 and on seabed: 1/2
Calculation range of depth	z = +3.5 to −8.5 m
Mesh sizes	Δx = 100 m (longshore direction) and Δz = 1 m (vertical direction)
Time intervals	Δt = 1 h
Boundary conditions	Left and right boundaries: $q_x = 0$ Landward and offshore boundaries: $q_z = 0$

Table 1.
Calculation conditions.

bathymetry in 1920 (**Figure A1**). During the first period, the parallel contours in 1899 simply advanced by 42 m until 1920, so that a new coordinate was taken for the shoreline to be shifted by 42 m seaward. During the second period, a V-shape shoreline recession began around the river mouth because of the decrease in Q_{in} from 4.7×10^5 m^3/year before 1920 to 1.5×10^5 m^3/year. However, the contour lines far

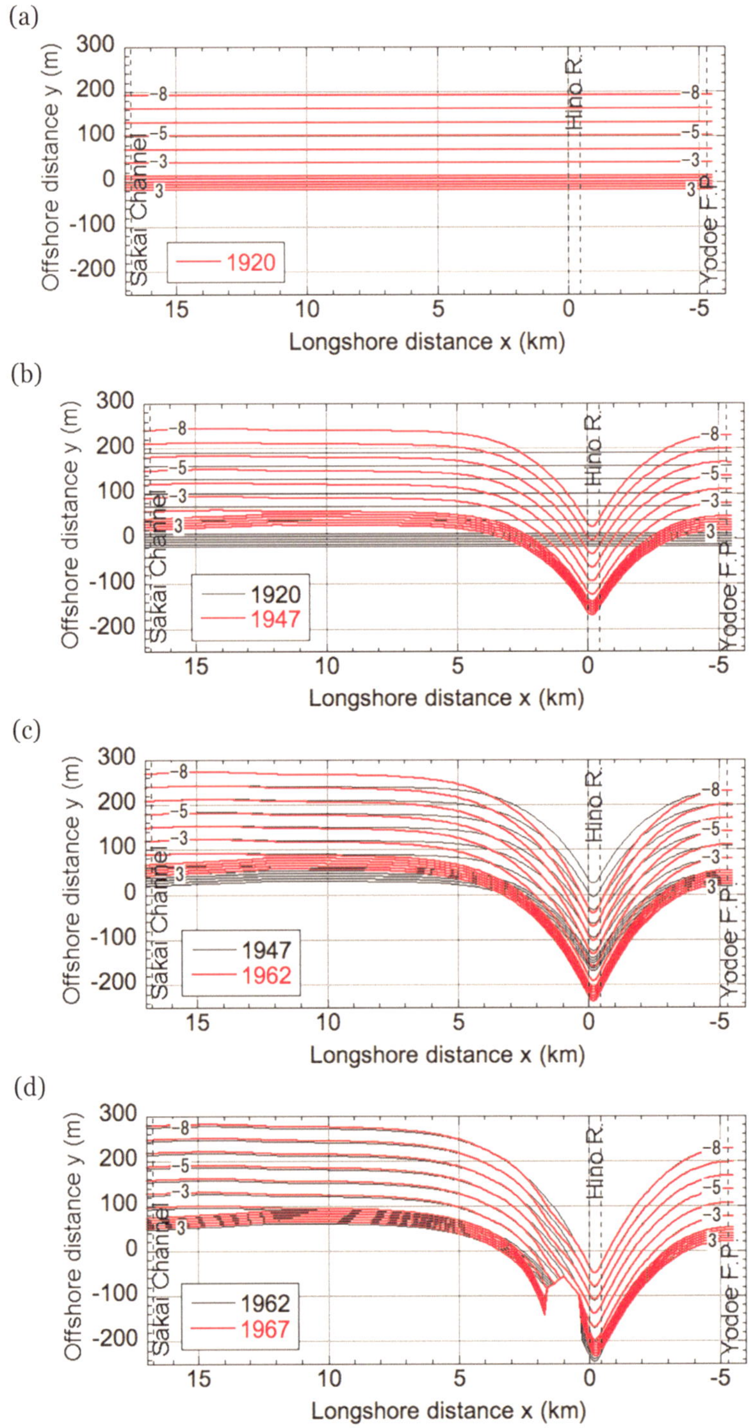

Figure 12.
Bathymetric changes from the first to fourth periods. (a) First period (1899–1920). (b) Second period (1920–1947). (c) Third period (1947–1962). (d) Fourth period (1962–1967).

from the river mouth continued to advance because of the successive supply of sand to both sides of the river mouth. During the third period, the shoreline markedly receded around the river mouth because of the further decrease in Q_{in} from 1.5×10^5 to 6×10^4 m^3/year owing to the stoppage of mining of iron sand. Even during this period, however, the contour lines continued to advance far from the river mouth. During the fourth period, a seawall was constructed west of the river mouth blocking westward longshore sand transport partially, so that the contour lines receded west of the seawall, even if Q_{in} was maintained at 6×10^4 m^3/year.

The shoreline changes in the entire period relative to that in 1920 can be drawn, as shown in **Figure 13**. The shoreline advanced by 42 m between 1899 and 1920 because of the accretion of the entire river delta. After 1920, however, the shoreline receded around the river delta owing to the decrease in sand supply, whereas the shoreline continued to advance far from the river mouth.

Figure 14 shows the measured and predicted shoreline changes with reference to the shoreline in 1899 and 1947. Regarding the shoreline changes until 1947, the

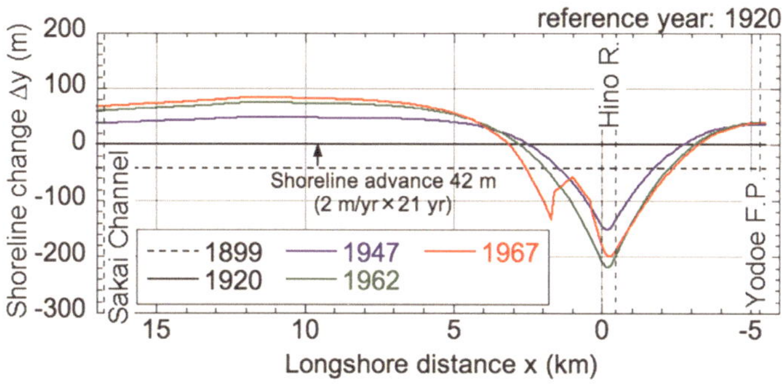

Figure 13.
Shoreline changes with reference to that in 1920.

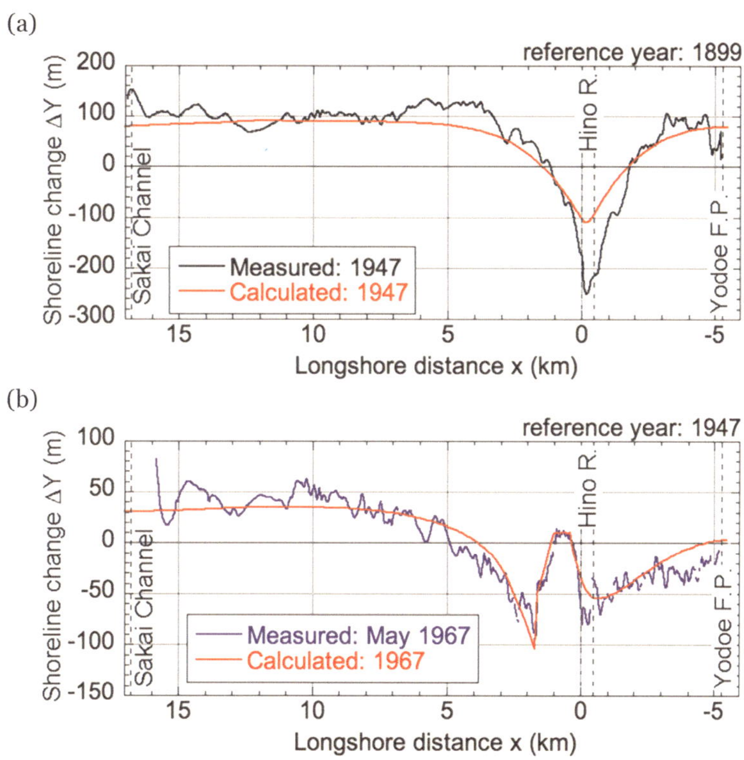

Figure 14.
Measured and predicted shoreline changes between 1899 and 1947 and between 1947 and 1967. (a) 1899–1947. (b) 1947–1967.

overall shoreline changes were reproduced well, except for the river mouth area where the shoreline change was underestimated. Regarding the shoreline changes until 1967, the shoreline changes in the overall area including the shoreline recession west of the seawall constructed immediately west of the river mouth were well reproduced. The predicted and measured shoreline changes are in good agreement.

6. Reproduction of topographic changes relative to topography in 1947

6.1 Calculation conditions

The calculation domain is the same as that in the previous section. In the calculation, the entire period between 1947 and 2010 was separated into five periods, and Q_{in} in each period was calculated from the volumetric change transformed from the shoreline changes measured during the period by multiplying the characteristic height of beach changes. In the first and second periods until 1962 and 1967, respectively, Q_{in} was given as 6×10^4, and 4.7×10^4 m^3/year in the third, fourth, and fifth periods between 1967 and 2010 with a decrease in Q_{in} by 20% after 1967.

For the initial topography, the results of the reproduction calculation until 1947 were employed, assuming parallel contour lines on the expanded coordinates. In the

Q_{in}	1 (1947–1962): 6×10^4 m^3/year 2 (1962–1967): 6×10^4 m^3/year, seawall between x = 0 and 1.5 km 3 (1967–1980): 4.7×10^4 m^3/year, detached breakwater Kaike fishing port, and seawall between x = 0 and − 3.5 km 4 (1980–1995): 4.7×10^4 m^3/year, detached breakwater between x = −3.5 and − 2.5 km, and between 3.5 and − 7 km, seawall between x = −5 and 0 km, and between x = 3.5 and 13 km 5 (1995–2010): 4.7×10^4 m^3/year, sand back pass of 3×10^4 m^3/year, in which sand is subtracted at x = 13 km and nourished at x = 8 km, L-shape groin between x = 7 and − 8 km, artificial reef and detached breakwater between x = −5 and 0 km
Incident wave angle	Initial breaker angle in 1899 minus inclination angle of shoreline reproduced in 1947
Coefficient of sand transport	Coefficient of Ozasa and Brampton's [18] term ξ = 3.24
Remarks	▪ Fluvial sand supply from Hino River: sand source distributed in depth zone between z = +3 and − 8 m ▪ Sand back pass by a rate of 3×10^4 m^3/year was given by a sink and source of sand between z = 0 and + 3 m ▪ Wave transmission coefficient K_t = 0.2 (detached breakwater) except K_t = 0.4 between x = −1.4 and − 0.5 km ▪ Artificial reef K_t = 0.8 ▪ Port breakwaters at Yodoe and Kaike fishing ports and Sakaiminato Marina K_t = 0.0 ▪ Coefficient of Ozasa and Brampton's [18] term ξ = 1.62: detached breakwaters between x = −1.4 and − 0.5 km

Table 2.
Calculation conditions.

calculation, the changes in contour lines between +3 and − 8 m were calculated, assuming the same seabed slopes of 1/6 between z = +3 and − 2 m, and 1/30 between z = −2 and − 8 m. The number of grain sizes (N), the grain sizes of fine and coarse sand, the initial contents of each grain size of μ_1 and μ_2, the equilibrium slope corresponding to each grain size, and the wave conditions are the same as those in the reproduction calculation. The longshore distribution of the initial breaker angle was given by subtracting the inclination angle of the shoreline reproduced in 1947 from the initial breaker angle in 1899. The change in wave field by the construction of structures was calculated using the angular spreading method for irregular waves. The wave transmission coefficient of detached breakwaters was set to K_t = 0.2 as shown in **Table 2**, except for the detached breakwaters with K_t = 0.4 placed between x = −1.4 and − 0.5 km, and K_t = 0.8 for artificial reefs.

6.2 Results of the reproduction calculations

6.2.1 Change in contour lines

Figure 15 shows the calculation results between the first and fifth periods together with an additional expression of bathymetric changes relative to the bathymetry in 1947 (**Figure A2**). In the first period (**Figure 15(a)**), erosion concentrated around the Hino River mouth with gradual accretion west of x = 4 km because of the decreased Q_{in} of 6 × 10^4 m^3/year. Since the calculation was carried out using the expanded coordinates set on the shoreline in 1947, the triangular shoreline recession area shown in **Figure 15(a)** corresponds to the recession of the protruded shoreline of a river delta. The shoreline receded around the river mouth, whereas the contour lines advanced west of x = 5 km.

In the second period (**Figure 15(b)**), the seawall had been constructed between x = 0 and 1.5 km, although Q_{in} was the same (6 × 10^4 m^3/year) as that in the first period. Owing to the construction of the seawall, westward longshore sand transport was partially blocked at the protruded seawall, resulting in erosion immediately west of the structure with the accretion upcoast. In the third period, Q_{in} decreased from 6 × 10^4 to 4.7 × 10^4 m^3/year together with the construction of 12 DBs and Kaike fishing port breakwaters (**Figure 15(c)**). Soon after the construction of DBs, cuspate forelands were formed behind the DBs. Simultaneously, severe erosion occurred downcoast of Kaike fishing port located at x = 3 km because of the decrease in the westward longshore sand transport.

In the fourth period, although Q_{in} was the same (4.7 × 10^4 m^3/year) as that in the third period, new DBs were constructed between x = −3.5 and − 2.5 km, and between 3.5 and 7 km, and further seawall was constructed between x = −5 and 0 km, and between x = 3.5 and 13 km, as shown in **Figure 15(d)**. At this stage, land reclamation was carried out at the west end at the shoreline, and the shoreline length was decreased by 4 km, resulting in the advance of all the contour lines. Although the shoreline east of the structures was stabilized by the construction of many coastal structures, erosion is severe downcoast.

In the fifth period between 1995 and 2010 (**Figure 15(e)**), Q_{in} was kept constant at 4.7 × 10^4 m^3/year, the same as that in the fourth period. For the sand back pass, sand was excavated at a rate of 3 × 10^4 m^3/year at x = 13 km, and the same amount was supplied at x = 8 km. In addition to this, an L-shaped groin was constructed at x = 7 km, and an artificial reef and DBs were constructed between x = −5 and 0 km. The effect of blocking longshore sand transport by the land reclamation reached upcoast, and the contour lines between x = 13 and 7 km became straight.

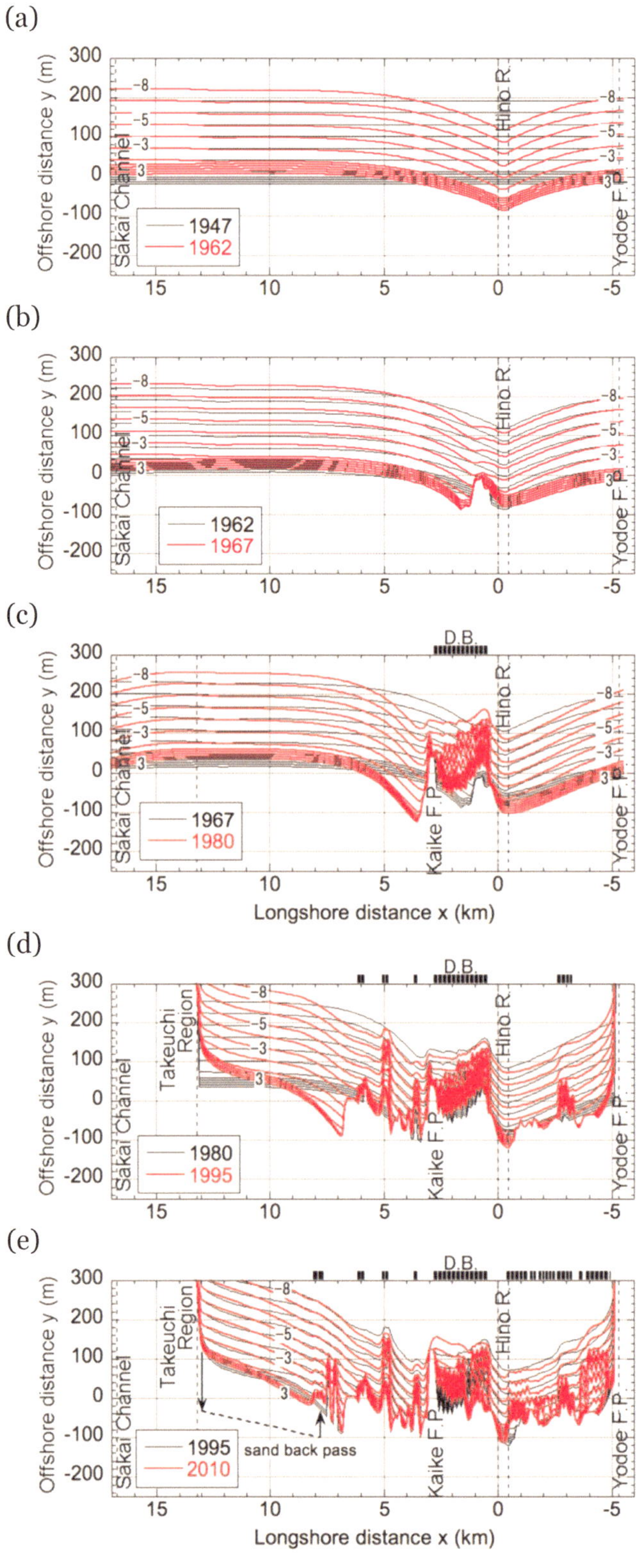

Figure 15.
Bathymetric changes between first period (1947–1967) and fifth period (1995–2010). (a) First period (1947–1962). (b) Second period (1962–1967). (c) Third period (1967–1980). (d) Fourth period (1980–1995). (e) Fifth period (1995–2010).

6.2.2 Depth changes

Figure 16 shows the measured and predicted shoreline changes between 1967 and 2007. The shoreline advance behind the DBs, downcoast shoreline recession, and the shoreline advance upcoast of the west boundary were in good agreement in the measured and predicted results. **Figure 17** shows the entire shoreline changes

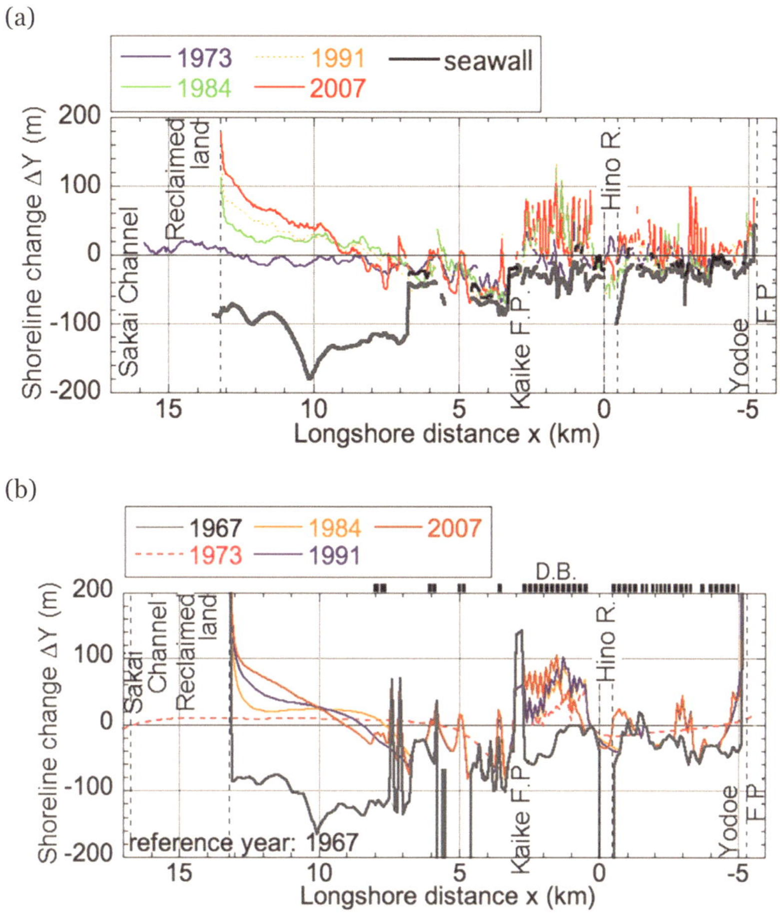

Figure 16.
Measured and predicted shoreline changes between 1967 and 2007. (a) Measured (1967–2007). (b) Calculated (1967–2007).

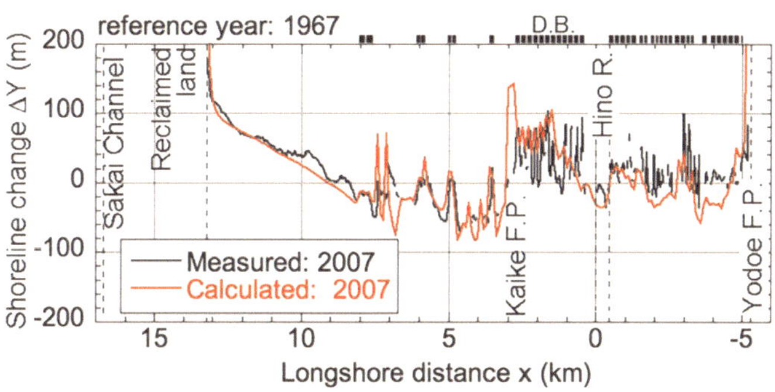

Figure 17.
Measured and predicted shoreline changes until 2007.

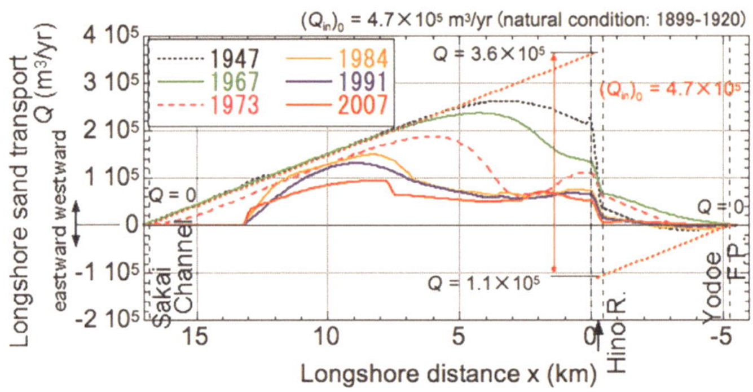

Figure 18.
Distribution of longshore sand transport rate.

until 2007 with reference to 1967. The predicted and measured shoreline changes agree well.

6.2.3 Comparison between measured and calculated shoreline changes

Figure 18 shows the distribution of longshore sand transport in the entire study area. Between 1899 and 1929, sand supplied from the Hino River was transported to both directions away from the river mouth with the westward and eastward transport of 3.6×10^5 and 1.1×10^5 m^3/year out of the entire sand supply of 4.7×10^5 m^3/year, respectively. With the decrease in sand supply from the Hino River, the longshore sand transport had decreased from the vicinity of the river mouth. In particular, the westward longshore sand transport markedly decreased after the construction of the DBs between 1967 and 1973 in the area between x = 0 and 3 km. East of the river mouth, eastward longshore sand transport at the initial stage reversed until 1967, and westward longshore sand transport began to occur after 1967.

7. Conclusions

The topographic changes of the Yumigahama Peninsula between 1899 and 2010 have been reproduced using the contour-line-change model considering the change in grain size of the seabed material. It was found that the beach changes of this peninsula were involved in the process leading to the reduction in the size of the Hino River delta, and strong erosion occurred around the river mouth. Because countermeasures were carried out from up the coast, sand was deposited up the coast of various structures with erosion down the coast. It was concluded that the contour-line-change model considering the change in grain size of the seabed material is a useful tool for predicting long-term topographic changes.

A. Appendix

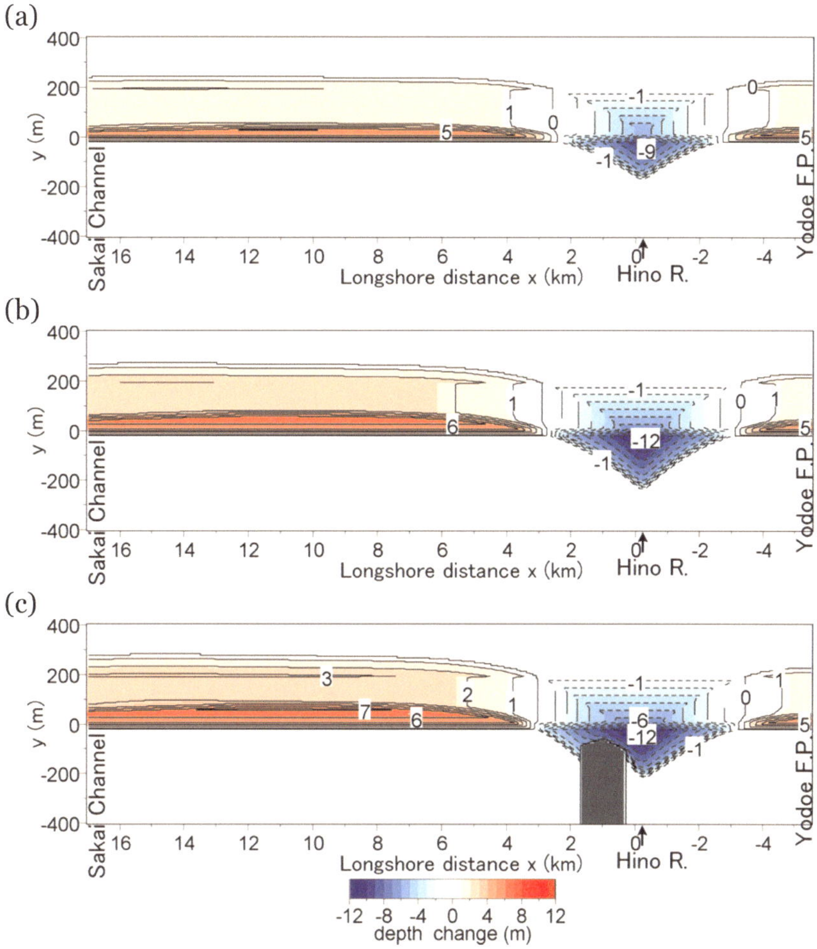

Figure A1.
Bathymetric changes until 1947, 1962, and 1967 with reference to bathymetry in 1920. (a) Bathymetric changes between 1920 and 1947. (b) Bathymetric changes between 1920 and 1962. (c) Bathymetric changes between 1920 and 1967.

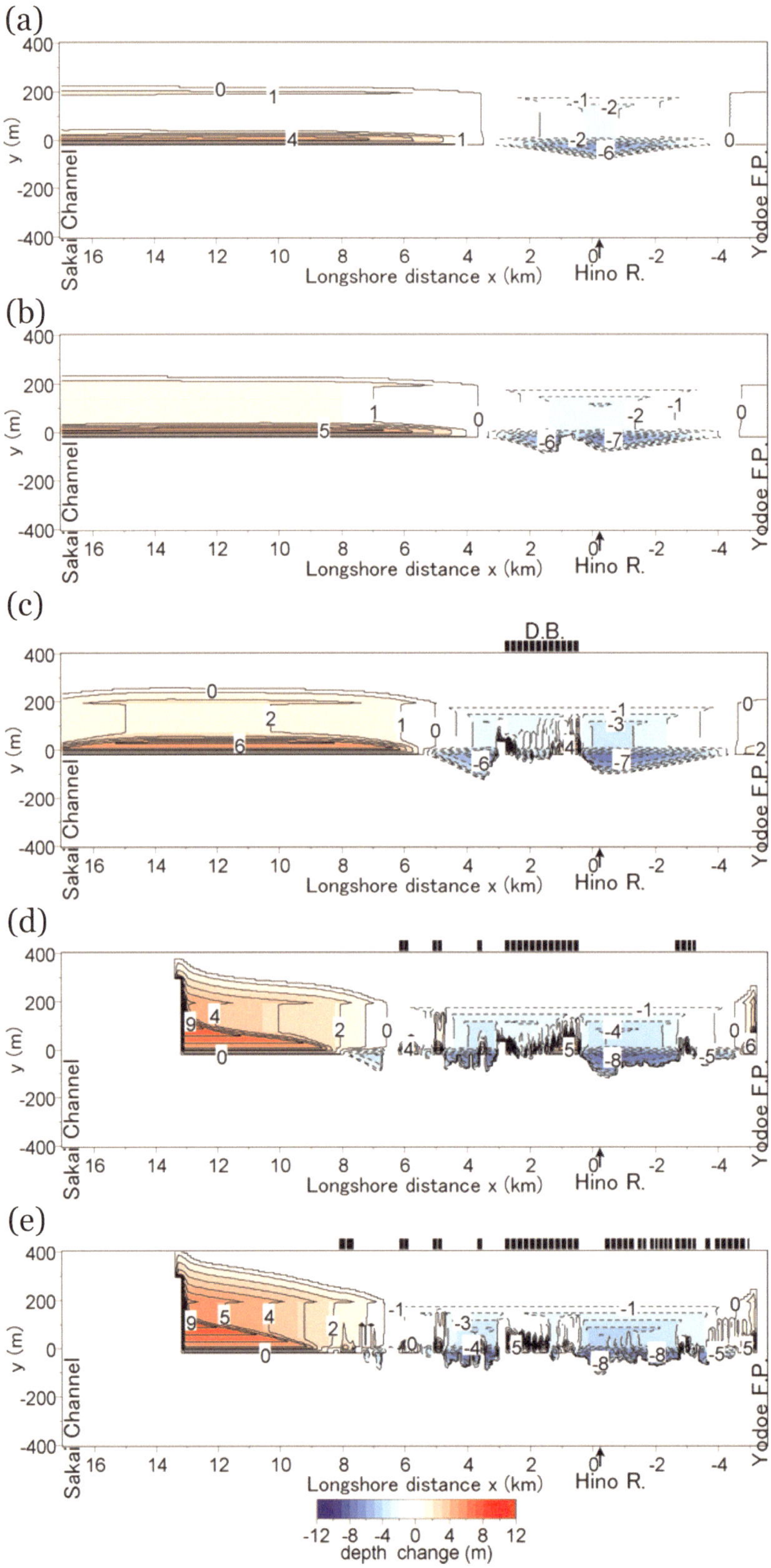

Figure A2.
Bathymetric changes with reference to bathymetry in 1947. (a) 1962. (b) 1967. (c) 1980. (d) 1995. (e) 2010.

Author details

Takaaki Uda[1], Shiho Miyahara[2]*, Toshiro San-nami[2] and Masumi Serizawa[2]

1 Public Works Research Center, Tokyo, Japan

2 Coastal Engineering Laboratory Co., Ltd., Tokyo, Japan

*Address all correspondence to: miyahara11@gmail.com

References

[1] Toyoshima O, Sadamichi N. Formation of tombolos and bathymetric changes on Kaike coast. In: Proceedings of the 21st Coastal Engineering Conference. JSCE; 1974. pp. 167-172 (in Japanese)

[2] Toyoshima O. Topographic changes owing to construction of detached breakwaters. In: Proceedings of the 24th Coastal Engineering Conference. JSCE; 1977. pp. 185-189 (in Japanese)

[3] Nishitani Y, Hayami M, Toyoshima O. Sand deposition effect of detached breakwater and bathymetric changes. In: Proceedings of the 28th Coastal Engineering Conference. JSCE; 1981. pp. 266-270 (in Japanese)

[4] Uda T. Beach Erosion in Japan. Tokyo: Sankaido Press; 1997. p. 442 (in Japanese)

[5] Uda T, Serizawa M, Miyahara S. Morphodynamic Model for Predicting Beach Changes Based on Bagnold's Concept and its Applications. London, UK: INTEC; 2018. p. 188. https://www.intechopen.com/books/morphodynamic-model-for-predicting-beach-changes-based-on-bagnold-s-concept-and-its-applications

[6] Bakker WT. The dynamics of a coast with a groyne system. In: Proceedings of the 11th Coastal Engineering Conference. ASCE; 1969. pp. 1001-1020

[7] Perlin M, Dean RG. A numerical model to simulate sediment transport in the vicinity of coastal structures. Miscellaneous Report, No. 83-10. Coastal Engineering Research Center, U.S. Army Corps of Engineers; 1983. p. 119

[8] Steetzel HJ, de Vroeg JH. Application of a multilayer approach for morphological modelling. In: Proceedings of the Coastal Sediments' 99. Vicksburg: ASCE; 1999. pp. 2206-2218

[9] Steetzel HJ, de Vroeg H, van Rijn LC, Stam JM. Long-term modelling of the Holland coast using a multi-layer model. In: Proceedings of the 27th International Conference on Coastal Engineering. Sydney, Australia: ASCE; 2000. pp. 2942-2955

[10] Dabees MA, Kamphuis JWNLINE. Efficient modeling of 3D beach change. In: Proceedings of the 27th International Conference on Coastal Engineering. Sydney, Australia: ASCE; 2000. pp. 2700-2713

[11] Hanson H, Larson M. Simulating coastal evolution using a new type N-line model. In: Proceedings of 27th International Conference on Coastal Engineering. Vicksburg: ASCE; 2000. pp. 2808-2821

[12] Hanson H, Aarninkhof S, Capobianco M, Jimenez JA, Larson M, Nicholls R, et al. Modelling coastal evolution on early to decadal time scales. Journal of Coastal Research. 2003;**19**:790-811

[13] Uda T, Serizawa M. Model for predicting topographic changes on coast composed of sand of mixed grain size and its applications (chap. 16). In: Angermann L, editor. Numerical Simulations—Examples and Applications in Computational Fluid Dynamics. Croatia: INTEC; 2010. pp. 327-358

[14] Uda T. Japan's Beach Erosion—Reality and Future Measures. 2nd ed. Singapore: World Scientific; 2017. p. 530

[15] Uda T, Yamagata H, Katoh K, Akamatsu N. Predictive model of three-dimensional development and deformation of a river mouth delta by applying contour line change model. In:

Proceedings of the 26th International Conference on Coastal Engineering. 1998. pp. 3138-3150

[16] Furuike K, Uda T, Serizawa M, San-Nami T, Ishikawa T. Model for predicting long-term beach changes originating from accretive features of a natural delta coast. In: Proceedings of the 5th International Conference on Asian and Pacific Coasts. Vol. 4. 2009. pp. 266-272

[17] Miyahara S, Uda T, Furuike K, Serizawa M, San-nami T, Ishikawa T. Effect of sand bypassing at Sakuma Dam in Tenryu River as a measure against erosion of Tenryu River delta coast. In: Proceedings of the 32nd ICCE, Sediment. 106. 2010. pp. 1-12. http://journals.tdl.org/ICCE/article/view/1049/pdf_165

[18] Miyahara S, Uda T, Ishikawa T, Furuike K, Serizawa M. Prediction of long-term topographic changes of Tenryu River delta associated with sand bypassing at dam in upper basin assuming no coastal facilities. In: Proceedings of the 6th International Conference, Asian and Pacific Coasts. 2011. pp. 216-223

[19] San-nami T, Furuike K, Uda T, Serizawa M. Formation of an arc-shaped accretive shoreline downcoast of sea cliffs and prediction of deformation. In: Proceeding of the Coastal Sediments. 2011. pp. 1243-1256

[20] San-nami T, Uda T, Ohashi N, Iwamoto H, Serizawa M, Ishikawa T, et al. Prediction of beach erosion caused by reduction of fluvial sand supply due to excess and mining and beach recovery after prohibition of mining. In: Proceedings of the 33rd International Conference on Coastal Engineering. Sediment. 61. 2012. pp. 1-11

[21] Kamphius JW, Davies MH, Narim RB, Sayao OJ. Calculation of littoral sand transport rate. Coastal Engineering. 1986;**10**:1-12

[22] Kumada T, Kobayashi A, Uda T, Serizawa M. Development of predictive model of shoreline and grain size changes. In: Proceedings of the Coastal Sediments. 2003. pp. 1-14

[23] Uda T, Kawano S. Development of a predictive model of contour line change due to waves. In: Proccedings of JSCE, No. 539/II-35. 1996. pp. 121-139 (in Japanese)

[24] Ozasa H, Brampton AH. Model for predicting the shoreline evolution of beaches backed by seawalls. Coastal Engineering. 1980;**4**:47-64

[25] Kraus NC. Field experiment on vertical mixing of sand in the surf zone. Sedimentary Petrology. 1985;**55**:3-14

Chapter 6

The Role of Mangroves in Coastal and Estuarine Sedimentary Accretion in Southeast Asia

Punarbasu Chaudhuri, Subhamita Chaudhuri and Raktima Ghosh

Abstract

Mangroves provide a distinctive mechanism of trapping sediment and accelerating land-building processes in tide-dominated coastal and estuarine environments. The complex hydrodynamic and salinity conditions, accumulation rates of both organic and inorganic sediments, primary surface elevation, and hydroperiod influence sediment retention mechanism within mangrove ecosystems. Abundant terrigenous sediment supply can form dynamic mud banks and the complex aerial root system of mangroves may lead to accretion of sediment by weakening the tidal velocity. Such mechanisms are often enhanced by organic flocculation. The efficiency of sediment trapping by mangroves is species specific. Adaptability and resilience of mangroves enable them to cope with the moderate to high rates of sea level rise. However, subsurface movements and deep subsidence due to autocompaction may augment the effects of relative sea level rise. Increasing population pressure and forest-based economic activities have caused global reduction of mangrove coverage challenging the sedimentation processes. Marker horizon techniques and surface elevation table (SET) tests have facilitated assessment of spatial variability in patterns of sediment accretion and surface elevation in various coastal sites of species-diverse Southeast Asia, especially coastal Malaysia and Thailand. The mangroves of the eastern coast of India have witnessed sediment retention, having an association with the seasonal rainfall regime.

Keywords: mangrove zonation, sediment accretion, estuarine, flocculation, turbidity maximum zone, bioturbation, propagules, autocompaction, sea-level rise, surface elevation, marker horizon

1. Introduction

Mangroves form the coastal and estuarine wetland ecosystem in the tropical and subtropical regions of the world. This unique intertidal ecosystem acts as a safeguard to the coastlines from the disastrous effects of storm surges, erosion, and floods. Some mangroves occur along open coasts, subject to moderate wave processes, while most of them grow in sheltered, muddy tracts that are either regularly or occasionally immersed by tides [1]. The extent and biomass of the

IntechOpen

mangrove forests are determined by variations in rainfall, tidal influence, wave energy, duration of tidal inundation, and salinity levels at both local and regional scale which further modify the physical and biological setup within a single coastal or estuarine area by affecting the water circulation pattern as well as sediment mixing and trapping [2]. There are 9 orders, 20 families, 22–27 genera, and roughly 70 species of mangrove, among which 40 exclusive and 65 non-exclusive species are found in Southeast Asia [3]. The zonation pattern of different species is attributed by salinity, tidal flooding, and land elevation. They develop in monospecific bands, parallel to the coastline. In Andaman and Nicobar island group, the mangroves depict a unique assemblage with the coral reefs. A general description of mangrove zonation extends from shore to the inland areas where at the edges the species are mostly salt-tolerant and at the interior parts the species are more adopted to nonsaline environment. There are four major hypotheses explaining the zonation—(i) land-building and plant succession hypothesis by Davis [4] elucidated the succession process by which the pioneer species of mangroves mold the substrate by trapping sediments in the intertidal zones and progressively mature mangrove species develop to compete with the colonizing species at established substrate; (ii) Woodroffe [5] has demonstrated the geomorphologic influences and long-term stratigraphic configuration to be dominant factors in species zonation; (iii) in 1980, Ball [6] has discovered the influence of salinity among other physicochemical determinants (seed dispersal, water logging, and tidal inundation) on species competition and growth which ascertain the forest structure; and (iv) Rabinowitz [7] laid emphasis on the propagule dispersal. Tidal action is responsible for delivering propagules of all sizes to their specific and suitable areas. Survival, establishment, and growth of propagules, therefore, play a significant role in dictating mangrove zonation. Predation of propagules by small crabs is correlated with the conspecific dominance and tree distribution. *Avicennia marina* is most heavily preyed upon, followed by *Ceriops tagal*, *Bruguiera gymnorrhiza*, and *Rhizophora stylosa* [2].

Woodroffe [9] outlined the relationship between the role of mangroves and the morphodynamic response of the shoreline. Sedimentation process including deposition of fine-grained, clay-dominant particles within the forest floor is considered to be one of the driving factors of land-building and shoreline progradation. Sedimentation modifies the geomorphological setup and influences the soil characteristics, groundwater reach, and substrate salinity determining mangrove zonation and species distribution. Mangrove vegetation favors the sedimentation process by resisting the tidal water flow and trapping the sediments through the network of their roots. The resistance offered by mangrove trees to water flow has been experimentally tested in a flume [8]. This makes the mangrove shorelines as remarkable sediment sink, characterized by long-term import of sediments, especially recent sediments which underlie the mangrove forests and coastal plains. The mangroves are mostly associated with muddy shorelines of the tropical deltas, but they may grow on a wide variety of substrates, including sand, volcanic lava, or carbonates. The carbonate sediments are derived from calcareous skeletal remains or coral reef substrates, but often the mangrove forest floor is underlain by organic peat, acquired mostly from mangrove roots. These sediments which are generated within the ecosystem are termed as in situ or autochthonous [9]. Allochthonous sediments are transported from the catchment through fluvial discharge or inflow of tidal current and littoral drift. The accumulation rates of allochthonous and autochthonous sediments, both inorganic and organic, differ between and within different geomorphological setups [9]. Dumped dredged material and other bottom sediments also contribute to allochthonous sediments when these are re-suspended by waves, turbulence generated by ships, and also dredging [10].

Country	Area ($\times 10^5$ ha)
Brunei	0.17
Cambodia	0.60
Indonesia	45.4
Malaysia	6.4
Myanmar	3.8
Philippines	1.6
Thailand	2.6
Vietnam	2.5
Total	63.2 (34.9% of the world)

Data from Spalding [11].

Table 1.
Mangrove areas in countries of Southeast Asia.

South and Southeast Asia represents some 42% of the total mangrove areas in the world and is typified by highest diversity of mangroves [11]. A large number of islands and a considerable total length of coastline permit mangrove growth and development in the coastal Southeast Asia. In most of the regions, coastlines are characterized by high rainfall and a large amount of riverine sediment input. The Ganges-Brahmaputra delta, constituted by a complex network of estuaries, tidal creeks, and islands, supports the world's largest continuous single-area mangroves—the Sundarbans [12]. The sediment input of this delta plain is mainly sourced by delivery from overbank flooding of the large rivers and their distributaries [13]. Almost 70% of the total mangroves of India exist in the deltaic region. Mahanadi delta constitutes the second most developed mangrove forests within India after the Indian Sundarbans [11]. Other mangrove areas of Southeast Asia include the Philippines, Brunei, Cambodia, Myanmar, Pakistan, Indonesia, Thailand, Malaysia, Singapore, Japan, China, and Vietnam (**Table 1**).

The goals of the chapter include (1) the review and reassessment of the sediment accretion pattern and processes of the Southeast Asian mangrove forests, (2) understanding the role of mangroves both as plants and ecosystem in accreting sediment in different geomorphic settings, and (3) relating surface elevation changes with relative sea-level rise patterns.

2. Tidal dynamics within mangrove forest

Estuarine circulation is often influenced by the asymmetry between the ebb tide and flood tide, mixing of saline and fresh water, and tidal range. The duration of the flood tidal current is of shorter span, but with stronger peak currents than the ebb tide in most of the extensive vegetated coastal wetlands [2]. The velocity of the tidal current is ultimately determined by the ratio of the forest area to waterway area and the slope from the tidal creek into the forest [14]. Sedimentation in the world's most extensive mangrove regions is a function of retarded flow velocity of the ebb tide due to the bottom friction generated by the mudflat and flow around tree trunks, roots, and pneumatophores [5] and thus directly related to the density of vegetation [15]. These flows are complex with eddies, jets, and stagnation zones.

Manning coefficient is a popular engineering parameter to measure the flow velocity in relation to friction within the forest. Thus,

$$U = \frac{1}{n} h^{2/3} I^{1/2} \quad (1)$$

where U is water flow velocity, n is spatially averaged Manning friction coefficient, h is water depth, and I is water surface slope [14].

Value of n is within the range of 0.025–0.035 in typical sandy channels. It is two to three times lesser in the muddy estuaries as it diminishes with the grain size [14]. But in the mangrove forest, the friction generated by vegetation becomes prominent along with the bottom friction which increases the value of n. Drag forces and trapping of water is another characteristic of mangrove forests. Some amount of this incoming and outgoing tidal water is temporarily retained within the forest floor, for being returned to the main channel later (**Figure 1**).

Demuren and Rodi [16] have observed that meanders create a secondary circulation which sorts the sediment according to size ranging from fine-grained mud to gravel. It is driven by stratification of flow by density or salinity and particle concentration. Due to the secondary circulation, fine-grained silt and clay are accumulated on the sloping banks, whereas sand and gravels remain on the bed. This secondary circulation along with tidal pumping promotes flocculation of sediments in the estuary, and these mechanisms altogether create mudbanks.

The drag force and delayed water flow is enhanced during the wet season due to the freshwater buoyancy, and it is slowed down during dry periods. Wolanski and Cassagne [17] have observed high evaporation rate, decline of freshwater input, and higher salinity accompanied by slower rate of tidal flushing during the dry seasons in Konkoure River delta in Guinea. In addition to this physical complexity of water circulation, some estuaries with mangrove dominance exhibit floating mangrove debris including propagules which are likely to be accumulated upstream. Hence, the presence of tree roots, animal burrows, mounds, and debris exert a drag force and resistance to water flow [2, 14].

The sediment transport by the attenuating tidal flow is largely managed by several interrelated processes such as (i) tidal pumping and mangrove tidal prism, (ii) secondary circulation, (iii) flocculation and trapping of small particles at the turbidity maximum zone (TMZ), and (iv) microbial production of humus. The relative

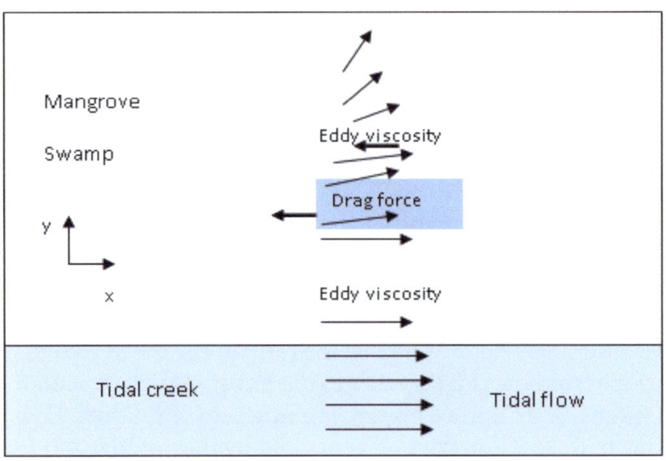

Figure 1.
Schematic view of the flow distribution and hydrodynamics in a mangrove swamp near a tidal creek, after Mazda et al. [15].

importance of these processes is site specific [2]. For example, an estuary with narrow fringe of mangrove has different tidal flushing characteristics and is unlikely that they play a significant role in sediment dynamics, whereas in the estuaries with extensive mangrove forest, sediment transportation and accretion are prominent.

Salinity stratification and salinity gradient determine the estuarine water circulation. In the presence of small freshwater input and small tidal range, a salinity maximum zone can develop which isolates the upper reaches of the estuary from the coastal edges creating an inverse circulation pattern, such as in Klong Ngao mangrove creek in Thailand. Reference [18–20] demonstrate the dynamics of an extensive, mangrove-fringed estuary in Malaysia which receives a large and sufficiently steady freshwater incursion. They found that the salinity stratification is strong during neap tide, but the system de-stratifies during spring tide. Water flooding the mangroves has low salinity at neap tide and is saltier at spring tide.

2.1 Flocculation and sediment trapping at the turbidity maximum zone (TMZ)

The fine-grained sediment particles, brought by the rivers or produced due to coastal erosion, are deposited predominantly at the vicinity of TMZ of the estuary as either individual grains or in aggregated (flocculated) form [21–23]. TMZ usually marks the landward limit reached by the saline water where the inward bottom flow meets the outward river flow, thus creating a shallow convergent water layer [2]. It encompasses a large variation in suspended particulate matter, which varies from 0.1 gl^{-1}, occurring at moderate to low freshwater flow situation, to more than 200 gl^{-1}, occurring at a prominent fluid mud layer with stationary suspension [20]. The turbidity maximum is not similar for all types of estuaries. It is largely controlled by degree of freshwater flow, salinity gradient, tidal dynamics, suspended particles at the upper reaches, etc. Researchers have attempted to investigate the pattern of sediment transport and characteristics of the turbidity maximum at different estuarine systems of the world through both laboratory-based and remote sensing-based methods. The degree of flocculation or colloidal stability [24] is largely dependent on a number of parameters including mineralogy [25], electrolytic levels which may alter with the changing salinity in estuary [26], organic content [27], suspended sediment concentration [28], and turbulent mixing [29, 30].

A cyclic occurrence of processes involved in sediment movement—suspension, flocculation, settling, deposition, erosion, and resuspension. Laboratory experiments revealed that flocculation occurs more readily when salinity increases [26]. However, salinity has an inverse relationship with settling velocity of the suspended particles. Laboratory analysis by Mhashhash et al. [31] reveals that settling velocity becomes faster with the increase in sediment concentration and decrease in salinity (**Figure 2**).

Cohesive sediments are composed of granular organic and mineral solids in a liquid phase [25]. In the estuaries cohesion of clay minerals is facilitated by the flat shape and size of particles with their surface area and electrical charge interacting with ambient water [32]. An important outcome of the cohesion property of sediments in the seawater is that particles can be adhered together and produce aggregates or flocs of several times greater than the size of the original or primary particles and can also be disaggregated. This reversibility between aggregation and disintegration of cohesive sediment is called flocculation [25, 33].

Flocculation leads to gentle mixing which increases the size from submicroscopic to microfloc, which grows in size after being merged with other microflocs [34]. The flocculated particles often provide surface area for absorbing heavy metal, pollutants, and nutrients. These processes control the size, density, and form of suspended particles [24, 35] and finally determine the settling velocity of flocs.

Once the floc gains its optimum size and strength, it is ready for sedimentation. Differential settling is the consequence of large particles with higher settling velocity colliding with smaller particles, having lower velocity [25, 33].

According to Winterwerp and Kesteren [25], a turbulent flow is responsible for introducing the particles into eddies, and the particles collide to produce flocs. Within the mangrove forests, turbulence is generated by flow around the trees, resulting in flocs which are composed of clay and silt particles. The settling of suspended sediment particles within the forest takes a shorter time (<30 minutes) during the transition from flood to ebb condition, when the water flow becomes relatively inactive [2]. Settling is also enabled by sticking of microbial mucus and by pelletization of invertebrate excreta. Mucus is not rare in mangroves, being found on rotting tree trunks and leaves, on the sediment surface, and in the density-driven lines of organic material [36] (**Figure 3**).

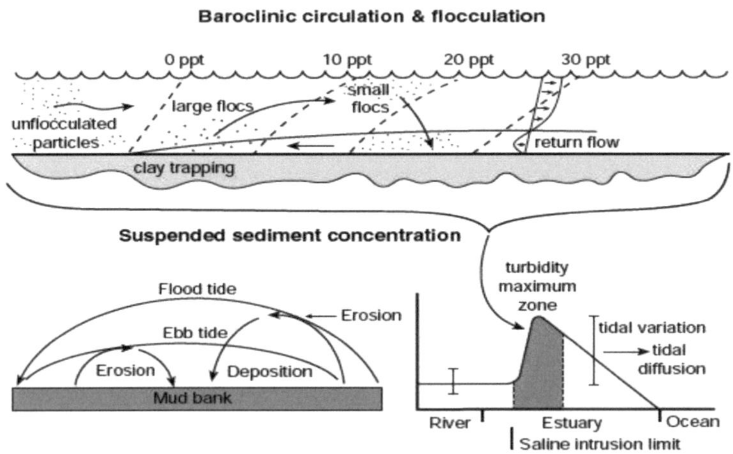

Figure 2.
The effect of baroclinic circulation, tidal pumping, mixing, and flocculation in the turbidity maximum zone of a model mangrove estuary, after Alongi [2].

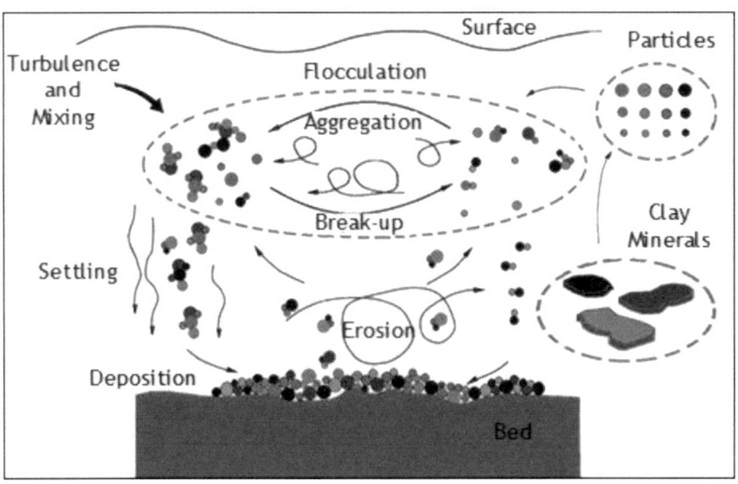

Figure 3.
Flocculation process, after Mandoza [32].

3. The role of mangrove in sedimentation

3.1 Sediment accretion among different root types

The most remarkable adjustments of the mangroves to the coastal and estuarine environment are the robust root structures which largely contribute to the geomorphological stability of the mudflats through trapping and binding of sediments [37]. The aboveground complex root structures of mangroves facilitate sediment accretion by increasing friction and reducing tidal current velocities [38]. According to Furukawa and Wolanski [14], mangrove forest acts as a "pump" of fine-grained sediments from coastal edges toward the forests. Pumping, here, indicates the turbulence generated by different mangrove root structures at the time when water enters into the forests as flood tide [39]. During slack tide the tidal current velocity slows down, ultimately becoming zero and resulting in deposition of flocs. The flocculated materials often grow in size and sometimes attain a size which the ebb tidal currents are unable to re-suspend [14].

The aboveground aerial roots are generally exposed in the tropical mangrove swamps [3]. However, the mangroves facing the waterfront have their aerial roots submerged during flood tides [3]. Tomlinson described "pneumatophores" as the upward extended erect root forms of the subterranean root systems. In *Avicennia* (**Figure 4b**), the pneumatophores are of limited height, commonly less than 30 cm. Cone roots also belong to pneumatophores and are developed by *Sonneratia* spp. and *Xylocarpus moluccensis*. The cone roots of *Sonneratia* (**Figure 4f**) attain greater heights (exceptionally up to 3 m) due to longer period of root development [3]. The size of pneumatophores of *Laguncularia racemosa* is of 20 cm in height. Root systems of *Avicennia* type (*Avicennia* species, *Sonneratia* species, and *Laguncularia racemosa*) offer the stability through their star-shaped network of cable roots radiating out from the trunk at a depth of 20–50 cm [40]. Apart from aerial roots (or pneumatophores) and cable roots, *Avicennia marina* possesses nutritive roots (or feeding roots) and anchor roots [41]. However, Spenceley [42] suggested that pneumatophores are likely to have better sediment-retaining properties than other root types. Buttress roots, developed mainly by *Heritiera littoralis* (**Figure 4d**) *and Pelliciera rhizophorae*, provide a strong tree stability in deltaic plain. Prop roots or stilt roots of *Rhizophora* spp. (**Figure 4a**) arch out from the tree trunk and often anchor within 30 cm of depth [43]. Stilt roots develop to a limited extent in *Bruguiera* and *Ceriops*. At the sapling stage, they grow at the stem base and form shallow buttresses in old trees [3]. *Ceriops* and *Bruguiera* possess aboveground knee roots (**Figure 4e**) with lenticels. Horizontally extended surface roots, developed by *Excoecaria agallocha* (**Figure 4c**), are prevalent in the tropical mangrove wetlands. Sedimentation potential around various mangrove species with differing cross-sectional root area is studied by Furukawa and Wolanski [14]. Species with prop roots, such as *Rhizophora* spp., tend to capture more sediments than *Ceriops* spp. which have smaller root knees. Moreover, the magnitude of sedimentation is greatest for trees forming a complex matrix of roots such as *Rhizophora* spp. and smallest for single trees like *Ceriops* spp. [14]. According to Scoffin [44] *Rhizophora* roots are the strongest binders of sediment as they reduce flow velocity to a degree which restricts sediment transportation. Specific root length and longevity of roots are other contributors to soil volume and thus to elevation gains [45]. Accumulation of long-lived roots through the loss, decomposition, and compression of cell contents often promotes the increase in soil volume [45] (**Figure 5**).

Mangrove seedling density induces sediment accretion [47]. Experiment carried out at Palakuda, Sri Lanka, by Kumara et al. [48] unfolded that accretion rates and aboveground biomass accumulation were highest among the highest density of planted *Rhizophora mucronata* seedlings for more than 3 years.

Figure 4.
(a) Rhizophora spp. with stilt roots; (b) pneumatophores of Avicennia spp.; (c) surface roots of Excoecaria spp.; (d) buttress root of Heritiera littoralis; (e) knee roots of Bruguiera gymnorhiza; (f) cone roots of Sonneratia alba; (g) root system exposed due to erosion of substrate; (h) discolored leaves, roots, and trunks indicating tidal submergence level; (i) thick muddy substrate due to sediment accretion. Photographs a, b, c, d, g, h, and i—Indian Sundarbans—are taken in November 2017 by Subhamita Chaudhuri. Photographs e and f—http://www.mangrove.at/mangrove_roots.html [46].

3.2 Spatial variability of sediment accretion

Within the coastal wetlands, sediment accretion pattern varies spatially. It negatively relates with the distance from mangrove edge [39]. Sedimentation is associated with the suspended sediment concentration during tidal inundation which reduces from seaward fringe to the scrub zone [14]. Reed [49] postulated that the front mangroves are situated at the slurry zone which is the depository of sediments both from the rivers and estuaries. Hence, sediment accretion is concentrated more at the tidally regulated front mangroves than back mangroves. Victor et al. [50] from their experiments at Ngerdorch and Ngerikiil estuaries of Micronesia have documented the efficiency of mangroves in trapping 44% of riverine fine-grained sediment. These estuaries are subject to high rates of sediment erosion resulting from land clearing and poor farming practices, and this erosion is largely affecting the growth of coral reefs. Through the radionuclide experiment at the Ganga-Brahmaputra delta, Allison and Kepple [51] show the decreasing pattern of sedimentation from inland to the shoreline, indicating that the sediment has its source at the marine side and it is introduced through tidal inundation, storm surges, and seasonal monsoon setup of sea level. This tidal delta plain accounts for

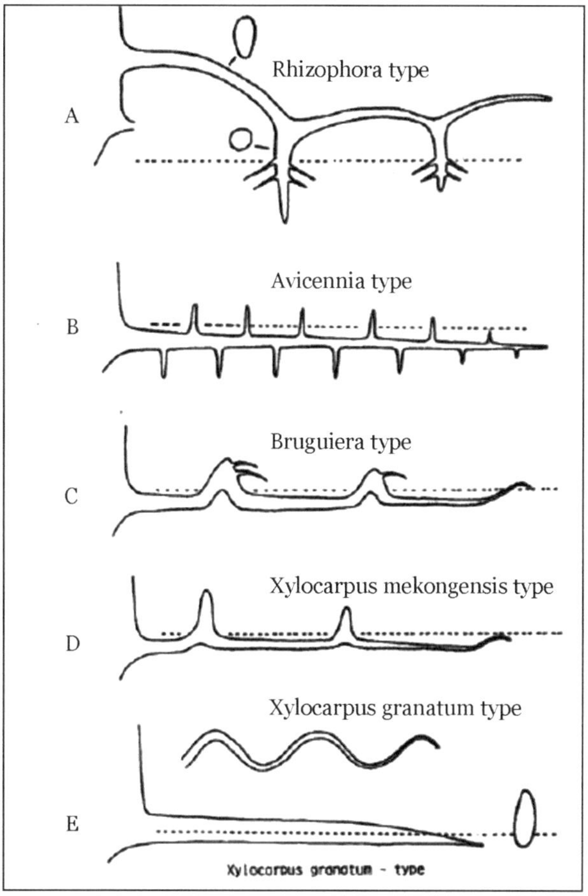

Figure 5.
Schematic diagram demonstrating aerial roots in mangroves, after Tomlinson [3]. All have developed from left to right. Dotted line represents substrate level.

a widespread mean annualized accretion rate of ~1.1 cm year^{-1}, and the heterogeneous sedimentation depositional pattern is influenced by tidal creek networks and topography [13, 51, 52]. However, researches of Santen et al. [53] demonstrate that in Red River, Vietnam, sediment accretion in some wave-dominated mangrove regions alternates with erosion in the fringe zones, and deposition mostly occurs in the riverbank mudflats.

Saad et al. [39], in their study at Kemaman River of Malaysia, found that the coarser and poorly sorted sediment accumulates at the front mangroves, while back mangroves are dominated by finer grains ranging from medium silt to very-fine-grained silt. The high energy waves at the front may lead to the deposition of medium sand at the front mangroves. The grain size often increases with the magnitude and frequency of storm surges [54].

Mangrove, as a community, influences sediment accretion in the different geomorphological units which are regularly modified by the physical forces and shoreline processes. Thom [55] has classified mangrove communities on the basis of their geomorphological setup as river-dominated, tide-dominated, wave-dominated and composite river-wave-dominated. The riverine mangrove areas, having a unidirectional flow, possess lesser sediment retention capacities. In contrast, tide-dominated fringes have the bi-directional flow facilitating net sediment import, sediment suspension and retention. Wave-dominated mangrove forests often possess distinctive sediment properties than others as because different bedform types promote strong erosion-accretion alternation within the forest [56].

Hydrogeomorphic variability along with variation in soil type in different topographic setups has led to the classification of mangroves as fringe, riverine, basin, scrub, and overwash. In Sri Lanka, these five types of mangrove systems are prominent [11]. Lugo and Snedaker [57] have demonstrated that dwarf mangroves exist in the environment with considerable scarcity of external nutrients. Both mineralogical and biological accretion are affected by the hydroperiod and complex morphodynamic feedbacks within these differentiated mangrove zones. The vertical elevation changes within these mangrove zones are not only the result of vertical accretion but also relate to the subsurface processes, such as compaction, decomposition, and shrink-swell cycles [58] (**Figure 6**).

In their studies at Southwestern Florida, Cahoon and Lynch [58] have observed that basin mangroves are often separated by berms, and the hydroperiod is mostly controlled by rainfall rather than tidal flushes except at extreme high-tide conditions. The accumulated sediment is mostly autochthonous, where organic matter inputs are prevalent and elevation changes are mainly caused by accretion and substrate shrink-swells due to cycles of flooding and drying. On the other hand, erosion and accretion processes are equally important for surface elevation changes in tide-dominated, fringe, and outwash mangroves [58] (**Figure 7**).

3.3 Seasonal change impacts on sediment accretion

Apart from the physicochemical properties present along the shoreline such as soil and water temperature, salinity, and pH, sedimentation processes are largely controlled by seasonal changes of the river discharge pattern and tidal regime. Due to notable increase in current velocity and river discharge, net sediment and organic matter transport rate progressively increase during the rainy months. At this time, buoyancy effect is important as the freshwater is captured in the forest during high tide [2]. Moreover, increasing erosion rates during wet seasons contribute to the

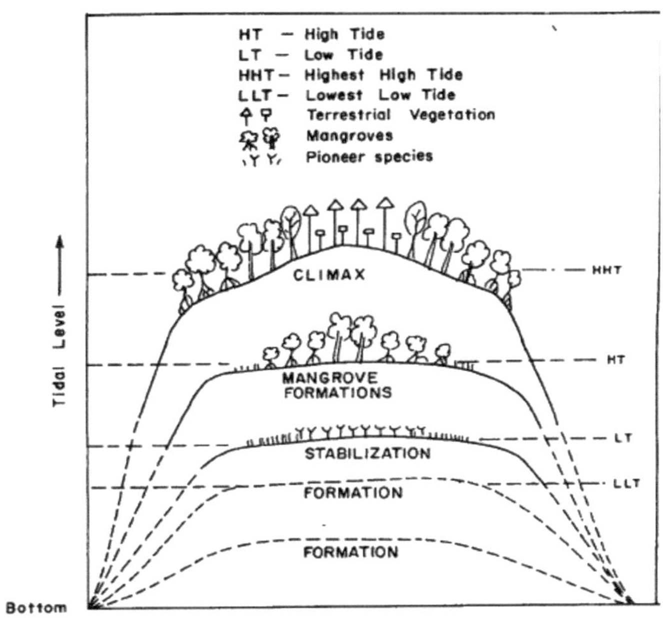

Figure 6.
Stages in the formation of mangroves in deltas, after Untawale and Jagtap [56].

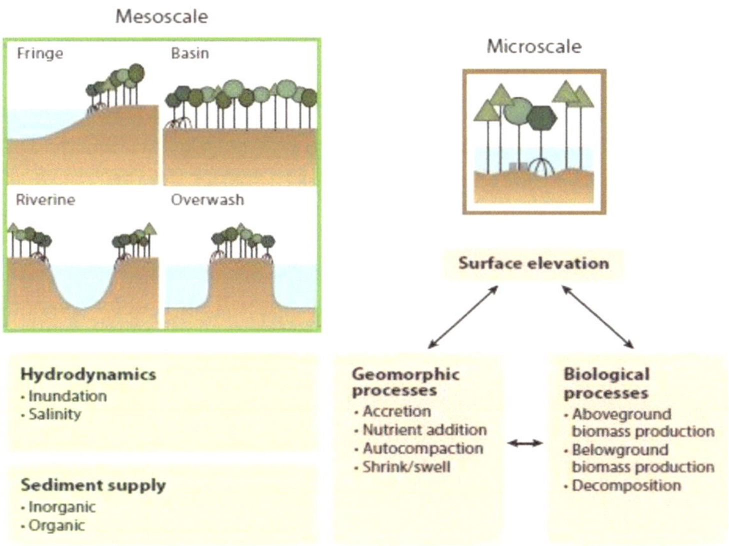

Figure 7.
Mesoscale processes represent interaction of mangroves mainly with hydrodynamics and sediment supply; microscale processes depicts mangrove stand interaction mainly with surface and subsurface processes, after Woodroffe et al. [1].

sediment budget at the foreshore. Relatively weak stratification of sediment is often recorded at the headwaters of mangrove-fringed waterways due to dominance of freshwater input during the wet season [2]. At spring tides, the ebb current directs the surface and bottom velocities for the whole tidal cycle in the upper estuary [59].

Degraded mangrove forests of the monsoon-influenced regions induce considerable sediment erosion which is ultimately followed by accumulation. Occasional storms often deliver the sediments within the forest and promote sediment deposition. Saad et al. [39] have observed the seasonal impact on sediment accretion rates in Kemaman, Malaysia, where the sediment accretion rate was 2.6 mm per month during the monsoons between November and January.

The low discharges during the dry season result in the landward transport of sediment. Residence time of water is long in the mangrove waterway during dry seasons. Trapping of water increases considerably as there is little freshwater to cause buoyancy-induced water circulation [2]. Spring tides often result in greater peak velocities at the surface in the middle or upper estuary. At the dry season, the saltwater reach extends to the extreme upstream section of the estuary. Reduced river discharge lowers the sediment input within the forest, resulting in slower rate of accretion. Saad's [39] observation at coastal Malaysia revealed that sediment concentrations reached only to 8–20 ppm in non-monsoon season as compared to 50–200 ppm during the monsoons. The average sediment accretion rate is eventually brought down to 1.2 mm per month during non-monsoon period.

3.4 Role of the ecosystem

Apart from diverse plant types, mangrove wetlands as an ecosystem support an incredible assemblage of fauna which, in turn, participate in land formation processes. The wide array of organisms includes barnacles, mollusks, shrimps, crabs, lobsters, jellyfish, tunicates, etc. which are often found among the roots of the mangroves. Autochthonous materials, including leaf litter, dead twigs, branches, and roots from the mangroves, accumulate on the mudflat surface and are incorporated within the soil through bioturbation by crabs [60]. This built-up material is consumed by detritivores, such as crabs, amphipods, and gastropod mollusks [61]. Some of this whole range of organisms plays a conspicuous role in aggregating and trapping sediments in their own way.

Mucus and bacterial populations are considerably abundant and productive in the mangroves. Mucus is known to be produced by benthic and pelagic detritivores. Wolanski [62] has noticed that the clay particles in suspension are trapped by sticking to the bacterial, algal, and animal mucus and pelletization by benthic detritivore deposits. He also found that a large number of non-flocculated particle, entering into the coral creek mangroves during flood tide, were re-exported at the ebb tide by sticking to the mucus floating on the surface water. This mucus was transported during ebb tide from the swamp to the creek where it produced prominent foam lines [62].

Animal structures, such as burrows, mounds, tubes, and other biogenic structures, also impact on sedimentation within the forest [63, 64]. These bioturbation structures are engineered by crabs and other benthic organisms. Numerous burrows generate friction on the forest floor when the tidal water flows through these burrows. Various models of fluid dynamics indicate that water circulation through the burrows are highly influenced by the architecture, slope, depth of the forest floor, location of roots relative to the burrow, and number of loops within the burrows [2]. De [65] in his experiments in the Indian Sundarbans demonstrated the biophysical mechanism of intertidal beach crab. Burrowing cycles within the substratum involve construction of oriented and open-to-air burrow tube (pre-tidal phase), formation of underwater and subsurface-trapped tabular air bubble occupied by the burrower that perpetually maintains internal and external pressure equilibrium by modifying inside burrow (tidal phase), and final exposure of air bubble system to air (post-tidal phase) before deserting the previous burrow and opening of another burrow cycle [65].

Vegetated marsh substrates reduce the fluid current velocity locally near the bed resulting in reduction of the energy available to move the sediment through fluid shear stress [39]. Benthic mats, developed by algal or microbial material on the surface of mangrove soils, contribute to vertical accretion [66]. These biomats are produced locally in patches in the depressed and moist areas on the supratidal flats following algal bloom. During the experiment at Hooghly estuary, De [65] observed that the mat grounds are cohesive, leathery, and composed of slightly coherent admixtures of fine-grained sand, silt, and green algae-secreted organic glue. Substrates of coastal wetlands including marshes and mangroves are thus characterized by organic matter deposition, suggesting the major role of biological processes in soil development, soil accretion, and elevation change [67–69]. Analysis of sediment cores has helped in inferring the contributions of organic matter to soil volume and vertical accretion in marsh and mangrove wetlands [70].

4. Sea-level changes, subsidence, and sediment accretion

Mangrove distribution along the coasts or estuaries changes with time, involving the balances between subsidence and accretion, erosion and vegetative stabilization, productivity and decomposition, tidal pumping, and drainage competency [1]. The global rise in sea level is caused by thermal expansion of seawater due to climate change and melting of polar ice caps and glaciers. These lead to the increase in volume of water in the ocean resulting in substantial rise in sea level, which is called eustatic sea-level rise. On the other hand, mean sea-level rise, measured by tide gauges, also varies because of tectonic movement, such as glacial-isostatic adjustments and lithospheric flexural subsidence [71]. Subsidence can be of two types: shallow and deep [72]. Shallow subsidence is caused by the reduction of mangrove and marsh surface elevation due to sediment compaction at the top layer, such as shrinkage of silt, clay, or peat deposits and accumulation of subsurface materials [73]. Deep subsidence is led by tectonic and isostatic processes. The measures for vertical accretion only consider the effect of shallow subsidence, whereas the methods measuring surface elevation anomalies include both shallow and deep subsidence [73]. The net effect of eustatic and isostatic sea-level changes results in the relative sea-level rise in a specific location over a specific time period [47]. Observed and projected sea-level rise has far reaching impacts on mangroves, from drowning the vulnerable wetlands to squeezing the coastal areas [74, 75]. However, satellite-based experiments of Phan et al. [74] at Mekong River showed that mangrove degradation and rapid coastal erosion has reduced the mangrove strip inducing lesser sedimentation. Krauss [45] showed that sediment accretion rates beneath some mangrove forests surpass the rates of sea-level rise. Hence, the subsurface processes play a dominant role in determining whether mangrove adjusts to sea-level rise. Subsidence led by autocompaction and areal expansion caused by mangrove root growth has important bearing on adjustment of mangroves to sea-level rise [1]. Surface elevation table measurements along with marker horizon techniques are often adopted to record vertical sediment accretion rates and substrate elevation changes and calculate short-term subsidence rates [1]. With the organic and mineral sediments, subsurface processes beneath the mangrove forests play a major role in developing surface elevation [76]. Wetland elevation is increased and inundation stress is decreased by sedimentation. Mangroves develop on the newly accumulated mudbank and facilitate soil development and elevation change [77]. Root growth dominates below-ground organic sediment accumulation, and this in turn keeps a balance with sea-level change [60, 78] (**Figure 8**).

The research findings of Cahoon and Lynch [58], based on mangrove forest of Southwestern Florida, have shown that vertical accretion is often driven by shallow subsidence and local sea-level rise. Hence, vulnerability of mangroves is described

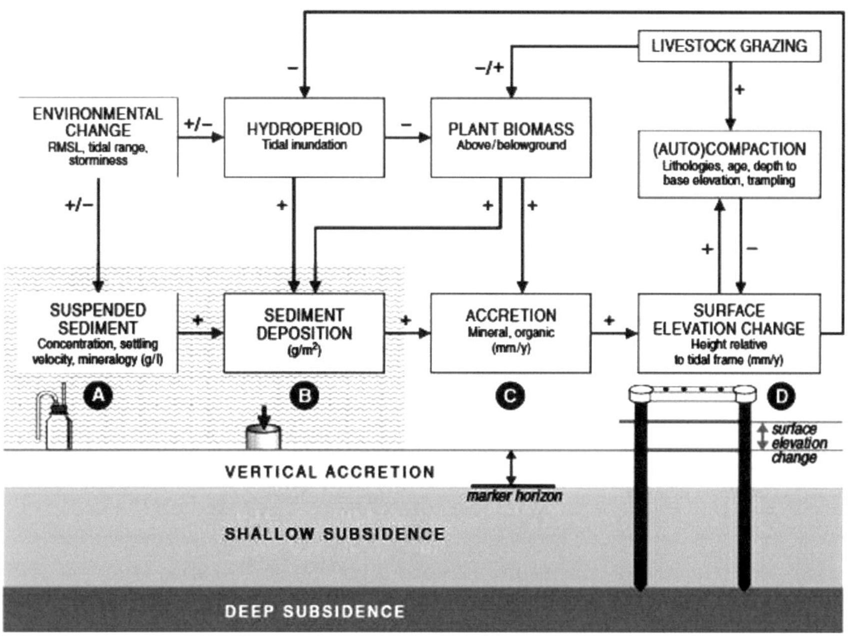

Figure 8.
Factors affecting sedimentation processes in coastal wetlands after Cahoon et al. [79].

Hydrogeomorphic setting	Surface elevation change (mm year^{-1})	Vertical accretion (mm year^{-1})	Subsurface change (mm year^{-1})
Fringe	−1.3 to +5.9	+1.6 to +8.6	−9.7 to +2.4
Riverine	+0.9 to +6.2	+6.5 to +13.0	−11.2 to −0.2
Basin	−3.7 to +3.9	+0.7 to +20.8	−19.9 to +2.8
Scrub	−1.1	−2.0	−3.1
Overwash	−0.6 to −2.5	+4.4 to +6.3	−3.8

Table 2.
Surface elevation change, vertical accretion, and subsurface adjustment for different mangrove hydrogeomorphic settings, determined using surface elevation table-marker horizon (SET-MH) methods, after Krauss et al. [45].

in terms of elevation reduction, rather than accretion deficit [58]. However, mangroves will be affected by inundation if the rate of sea-level rise is greater than vertical land development due to sediment accretion and root accumulation. In such situations, mangroves would naturally have the tendency to extend landward [1].

The mangroves respond differently to surface accretion, subsidence, and sea-level rise according to their hydrogeomorphic setting. In the experiment at Micronesian mangroves, Krauss et al. [37] found that the fringe and riverine mangroves are moderately susceptible to local sea-level rise, despite considerably high sediment accretion rates along Yela and Utwe rivers of Pacific high islands. In contrast, root and peat-based accumulation resulted in greater elevation gain in Belize [78]. Fringe mangroves are specifically vulnerable to sea-level changes than riverine or interior mangroves, partly due to physiological stress imparted by prolonged flooding [37].

According to Cahoon et al. [60], mangrove forests of the world are prone to lose surface elevation relative to sea-level rise, despite their ability to accrete sediment in some hydrogeomorphic settings. Hence, protecting the mangroves susceptible to sea-level rise in the outer margin of the estuaries from human interferences may slow the rate of soil loss [37] (**Table 2**).

5. Methodologies for measuring sediment accretion

Various methods and models have been adopted till date to analyze and understand sediment accretion rates within mangrove ecosystems of the world. Sedimentation rates, measured by short-term measurement of changes in relative sea level along with the estimates by radiotracers, provide a net sedimentation pattern [2]. There is a widespread use of radioisotopes ^{210}Pb and ^{137}Cs in analysis of long-term sedimentation within mangroves as well as salt marsh areas. Mudd et al. [80] have used the above method along with OIMAS-N model simulating the ephemeral evolution of a sediment column situated within a salt marsh [80]. Banerjee et al. [81] established the ^{210}Pb geochronology in selected four sediment cores in the Sundarbans and the Hugli estuary of India for the assessment of trace metal distribution in the sediment. The core sites were selected on the basis of different anthropogenic and hydrological parameters. Fe-Mn oxyhydroxide is observed to be the major controlling factor for trace metal accumulation as compared to organic carbon in both the sites. Fe-normalized enrichment factors (EFs) were calculated based on trace element abundance, and the result shows EF >1 for Cd, Pb, Co, and Cu indicating high enrichment in the top layers of the forest substrate. This is mostly because the core sites receive high pollution load from various anthropogenic sources. The experiment revealed relatively less contamination in the Hugli estuary due to high energy conditions and mostly coarse-grained sediments. Chaudhuri et al. [82] determined the accumulation of various trace metals in fine nutritive roots of *Avicennia marina* under the contaminated sediments at Sydney estuary, Australia. The result showed highest metal concentrations of Cu, Pb, and Zn in the fine nutritive roots, with a mean of 153, 189, and 378 mg/kg, followed by As, Cr, and Ni with mean enrichment concentration of 16, 21, and 11 mg/kg, respectively. Cd and Cr have low concentrations in fine nutritive root tissues. Radiocarbon (^{14}C) dating is another widely used method for measuring long-term sedimentation rates for both freshwater and marine ecosystems [83]. Optically stimulated luminescence (OSL) technique is often used for dating both older and younger (<60 years) sediments of the coasts. Madsen and Murray [84] have provided a detailed analysis and review on this technique.

"Marker horizon" technique and sedimentation plates are two most popular methods of quantifying transient sediment accretion [37, 73, 85]. Krauss et al. [85] in their study used sediment pins to measure elevation change in Micronesian mangrove forests. A marker horizon, mainly consisting of degradable material as opposed to sedimentation plates, assists as a reference layer within the soil, against which deposition of both mineral and organic sediment can be measured using a soil corer [72, 85]. In the case of dense vegetation, clipping the vegetation before applying the marker material is advisable, whereas marker material can be placed at soil surface within the stems in less dense forests. Following this procedure the effect of vegetation canopy structure on sediment accretion rates can be investigated [73]. With stacked layers of the markers, autocompaction rates are often assessed [73]. However, recovery of markers may be a challenging task when the layer is disturbed by bioturbation [84], distributed by profound floods, or shuffled with darker inorganic and organic materials [68]. In the sedimentation plate method, the marker horizon consisting of a firm plate made of metal or plastic is buried horizontally within the soil just below the rooting zone [73]. For recording sediment accretion, a thin metal pin is pushed within the soil until it reaches the plate, and its length above the sediment is determined [73]. However, measurements, based on marker horizon and surface elevation tables, provide short-time perspectives, indicating shallow subsurface processes of root growth and substrate autocompaction, whereas radiometric dating measures long-term sedimentation [1]. Saad et al. [39] applied the methodology based on estimating the thickness of a sediment section divided by the time span necessary for its deposition. The study covered a span of 2 years. To estimate the amount of sediment trapped by mangroves in the Ngerdorch and the Ngerikiil estuaries of Micronesia, Victor et al. [50] measured the salinity, temperature, and suspended sediment concentration to finally quantify the freshwater flow, brackish water outflow, and net estuarine sediment transport. Backscattering nephelometer is used widely by the scientists for estimating suspended sediment concentration. Horstman et al. [86] estimated dry weight of deposited sediments on the forest floor in the Andaman coast of Southern Thailand. They used ceramic tiles covered with smooth layer of mortar and carefully leveled with the forest floor for recording sediment entrainment and deposition. Computer-aided modeling of sediment transport is a valuable tool to understand and predict morphological change and sedimentation amount [87]. Delft3D software is very popular and is applied widely for simulating hydrodynamics, sediment dynamics, morphological processes, and biotic impacts in shallow water environments [88]. The process-based Delft3D-FLOW module solves the three-dimensional and two-dimensional unsteady shallow water equations. The hydrodynamic model applies horizontal momentum equations to compute transport and deposition of sediments concomitantly with the hydrodynamics,

Location	Sedimentation rate (mm year^{-1})
Bay of Bengal	≤5
Ajkwa estuary, Papua	0.6–5.5
Sawi Bay, Thailand	10–12
Matang Mangrove Forest Reserve, Malaysia	10–31
Kuala Kemaman Forest Reserve, Malaysia	10.6
Jiulongjiang estuary, China	13–60

Data from [39, 58, 78, 90–92].

Table 3.
Sedimentation rates in some mangrove forests focusing on Southeast Asia.

facilitating the understanding of morphodynamic feedback [86]. For measuring sedimentation rate and ratio, Adame et al. [89] used sediment traps which consisted of pre-weighed 9-cm Whatman qualitative filters placed in the ground over Petri dish lids held to the sediment by hooks. In addition, for the assessment of sediment quantity transported and deposited in the mangroves, they used glomalin—a novel terrestrial soil carbon tracer (**Table 3**).

6. Conclusion

Mangroves, acting as traps for both mineral and organic sediments, control the sedimentation and thus form their own survival ground. In the investigation at the Gulf of Thailand, Thampanya et al. [93] substantially differentiated coasts with and without mangroves. Coasts with mangroves showed prograding characteristics with low rate of erosion. The eroding coastal stretches are characterized by the absence of mangroves coupled with increased number of shrimp farms, increased fetch to prevailing monsoon, and decreased riverine inputs due to construction of dam [93]. Mangrove swamps of Southeast Asia are typified by sediment transport and circulation which is the consequence of intense anthropogenic disturbances near and around coastal regions and high rate of sediment erosion. Mangrove degradation and fluvial discharge with seasonal maxima seems to play a pivotal role in sediment erosion. Mangroves of Mekong delta have been especially affected by human activities including cutting of trees for timber and reclamation for shrimp cultivation [94]. Mandai mangroves, a small mangrove patch of Northeast Singapore, indicate an impact of urbanization. It has been a hotspot of research agenda for decades providing a broader context of Southeast Asian mangrove conservation [95]. Hence programs for plantation, restoration, and rehabilitation can alter the decline of mangrove habitat if proper hydrodynamics and sedimentary requisitions are met [1]. Alongi [96] examined the impact of climate change on mangrove forests. The Intergovernmental Panel on Climate Change (IPCC) has predicted that mangroves of arid coastlines, in subsiding river deltas, and some islands will reduce in area, though they have often proved to be either resilient or resistant to most environmental changes. The persistence of mangroves insinuates their ability to cope with moderately high rates of relative sea-level rise [1]. High sediment accretion, coupled with surface elevation change and plant survival in high densities, can facilitate shoreline protection and counter relative sea-level rise in the tropics. Moreover, continuous increment of aboveground biomass within the high-density mangrove wetlands not only advocates surface elevation gain but also acts as atmospheric carbon sink [48]. As they grow in saturated, muddy, low-oxygen soils, maximum amount of carbon is stored in roots, resists decay, and becomes long-term sinks as mangrove peat [61]. Thus, they provide other significant ecological services such as carbon storage. These tidal forests of the tropics are unique open ecosystems for a variety of structural and functional properties as well as their distinctive adaption techniques with the hydrogeomorphic processes.

Author details

Punarbasu Chaudhuri[1*], Subhamita Chaudhuri[2] and Raktima Ghosh[3]

1 Department of Environmental Science, University of Calcutta, Ballygunge, Kolkata, West Bengal

2 Department of Geography, West Bengal State University, Barasat, Kolkata, West Bengal

3 Department of Geography, University of Calcutta, Ballygunge, Kolkata, West Bengal

*Address all correspondence to: punarbasu_c@yahoo.com

References

[1] Woodroffe CD, Rogers K, McKee KL, Lovelock CE, Mendelssohn IA, Saintilan N. Mangrove sedimentation and response to relative sea-level rise. Annual Review of Marine Science. 2016;**8**:243-266

[2] Alongi D. The Energetics of Mangrove Fo\rests. Germany: Springer Science & Business Media; 2009

[3] Tomlinson PB. The Botany of Mangroves. Cambridge, UK: Cambridge University Press; 2016

[4] Davis JH. The ecology and geologic role of mangroves in Florida. Publications of the Carnegie Institution of Washington. 1940;**517**:303-412

[5] Woodroffe CD. Geomorphology and development of mangrove swamps, Grand Cayman Island, West Indies. Bulletin of Marine Science. 1982;**32**(2):381-398

[6] Ball MC. Patterns of secondary succession in a mangrove forest of southern Florida. Oecologia. 1980;**44**(2):226-235

[7] Rabinowitz D. Dispersal properties of mangrove propagules. Biotropica. 1978;**10**(1):47-57

[8] Struve J, Falconer RA, Wu Y. Influence of model mangrove trees on the hydrodynamics in a flume. Estuarine, Coastal and Shelf Science. 2003;**58**(1):163-171

[9] Woodroffe C. Mangrove sediments and geomorphology. Tropical Mangrove Ecosystems. Coastal and Estuarine Studies. 1992;**41**:Chapter 2

[10] Wolanski E. Physical Oceanographic Processes of the Great Barrier Reef. Florida: CRC Press; 1994

[11] Spalding M. World Atlas of Mangroves. Abingdon-on-Thames, UK: Routledge; 2010

[12] Stanley DJ, Hait AK. Deltas, radiocarbon dating, and measurements of sediment storage and subsidence. Geology. 2000;**28**(4):295-298

[13] Rogers KG, Goodbred SL Jr, Mondal DR. Monsoon sedimentation on the 'abandoned' tide-influenced Ganges–Brahmaputra delta plain. Estuarine, Coastal and Shelf Science. 2013;**131**:297-309

[14] Furukawa K, Wolanski E. Sedimentation in mangrove forests. Mangroves and Salt Marshes. 1996;**1**(1):3-10

[15] Mazda Y, Kobashi D, Okada S. Tidal-scale hydrodynamics within mangrove swamps. Wetlands Ecology and Management. 2005;**13**(6):647-655

[16] Demuren AO, Rodi W. Calculation of flow and pollutant dispersion in meandering channels. Journal of Fluid Mechanics. 1986;**172**:63-92

[17] Wolanski E, Cassagne B. Salinity intrusion and rice farming in the mangrove-fringed Konkoure River delta, Guinea. Wetlands Ecology and Management. 2000;**8**(1):29-36

[18] Wattayakorn G, Wolanski E, Kjerfve B. Mixing, trapping and outwelling in the Klong Ngao mangrove swamp, Thailand. Estuarine, Coastal and Shelf Science. 1990;**31**(5):667-688

[19] Ridd PV, Wolanski E, Mazda Y. Longitudinal diffusion in mangrove-fringed tidal creeks. Estuarine, Coastal and Shelf Science. 1990;**31**(5):541-554

[20] Uncles RJ, Stephens JA, Law DJ. Turbidity maximum in the macrotidal, highly turbid Humber Estuary, UK: Flocs, fluid mud, stationary suspensions and tidal bores. Estuarine, Coastal and Shelf Science. 2006;**67**(1-2):30-52

[21] Wu J, Liu JT, Wang X. Sediment trapping of turbidity maxima in the Changjiang estuary. Marine Geology. 2012;**303**:14-25

[22] Uncles RJ, Stephens JA. Turbidity and sediment transport in a muddy sub-estuary. Estuarine, Coastal and Shelf Science. 2010;**87**(2):213-224

[23] Woodruff JD, Geyer WR, Sommerfield CK, Driscoll NW. Seasonal variation of sediment deposition in the Hudson River Estuary. Marine Geology. 2001;**179**(1-2):105-119

[24] van Leussen W. Estuarine macroflocs and their role in fine-grained sediment transport [PhD thesis]. University of Utrecht; 1994

[25] Winterwerp JC, Van Kesteren WG. Introduction to the Physics of Cohesive Sediment Dynamics in the Marine Environment. Vol. 56. Amsterdam, Netherlands: Elsevier; 2004

[26] Krone RB. A study of rheologic properties of estuarial sediments (No. SERL-63-8). California University Berkeley Sanitary Engineering Research Lab; 1963

[27] Kranck K. The role of flocculation in the filtering of particulate matter in estuaries. In: The Estuary as a Filter. Cambridge, UK: Academic Press; 1984. pp. 159-175

[28] Burban PY, Lick W, Lick J. The flocculation of fine-grained sediments in estuarine waters. Journal of Geophysical Research, Oceans. 1989;**94**(C6):8323-8330

[29] Winterwerp JC. A simple model for turbulence induced flocculation of cohesive sediment. Journal of Hydraulic Research. 1998;**36**(3):309-326

[30] Manning AJ. The observed effects of turbulence on estuarine flocculation. Journal of Coastal Research. 2004;**41**:90-104

[31] Mhashhash A, Bockelmann-Evans B, Pan S. Effect of hydrodynamics factors on sediment flocculation processes in estuaries. Journal of Soils and Sediments. 2018;**18**(10):3094-3103

[32] Ramırez-Mendoza R. Flocculation Controls in a Hypertidal Estuary. [PhD thesis]. University of Liverpool; 19 April 2015

[33] Dyer K. Coastal and Estuarine Sediment Dynamics. Chichester, Sussex (UK): John Wiley and Sons; 1986. 358 p

[34] Eisma D. Intertidal Deposits: River Mouths, Tidal Flats, and Coastal Lagoons. Vol. 16. Florida: CRC Press; 1998

[35] Gibbs RJ. Estuarine flocs: Their size, settling velocity and density. Journal of Geophysical Research, Oceans. 1985;**90**(C2):3249-3251

[36] Stieglitz T, Ridd PV. Trapping of mangrove propagules due to density-driven secondary circulation in the Normanby River estuary, NE Australia. Marine Ecology Progress Series. 2001;**211**:131-142

[37] Krauss KW, Allen JA, Cahoon DR. Differential rates of vertical accretion and elevation change among aerial root types in Micronesian mangrove forests. Estuarine, Coastal and Shelf Science. 2003;**56**(2):251-259

[38] Wolanski E, Mazda Y, Ridd P. Mangrove hydrodynamics. Tropical mangrove ecosystems. Coastal and Estuarine Studies. 1992;**41**:Chapter 3

[39] Saad S, Husain ML, Yaacob R, Asano T. Sediment accretion and variability of sedimentological characteristics of a tropical estuarine mangrove: Kemaman, Terengganu, Malaysia. Mangroves and Salt Marshes. 1999;**3**(1):51-58

[40] Perillo GM. Geomorphology and sedimentology of estuaries: An introduction. In: Developments in Sedimentology. Vol. 53. Amsterdam, Netherlands: Elsevier; 1995. pp. 1-16

[41] Purnobasuki H, Suzuki M. Functional anatomy of air conducting network on the pneumatophores of a mangrove plant, Avicennia marina (Forsk.) Vierh. Asian Journal of Plant Sciences. 2005;**4**(4):334-347

[42] Spenceley AP. Sedimentation patterns in a mangal on magnetic Island near Townsville, North Queensland, Australia. Singapore Journal of Tropical Geography. 1982;**3**(1):100-107

[43] Macnae W. A general account of the fauna and flora of mangrove swamps and forests in the Indo-West-Pacific region. In Advances in marine biology. Academic Press; 1 Jan 1968. Vol. 6, pp. 73-270

[44] Scoffin TP. The trapping and binding of subtidal carbonate sediments by marine vegetation in Bimini Lagoon, Bahamas. Journal of Sedimentary Research. 1970;**40**(1):249-273

[45] Krauss KW, McKee KL, Lovelock CE, Cahoon DR, Saintilan N, Reef R, et al. How mangrove forests adjust to rising sea-level. New Phytologist. 2014;**202**(1):19-34

[46] Mangrove roots. Mangroves. Available from: http://www.mangrove.at/mangrove_roots.html [Accessed: 03 March 2019]

[47] McIvor AL, Spencer T, Möller I, Spalding M. The response of mangrove soil surface elevation to sea-level rise. Natural Coastal Protection Series: Report 3. Cambridge Coastal Research Unit Working Paper 42; 2013. ISSN: 2050-7941

[48] Kumara MP, Jayatissa LP, Krauss KW, Phillips DH, Huxham M. High mangrove density enhances surface accretion, surface elevation change, and tree survival in coastal areas susceptible to sea-level rise. Oecologia. 2010;**164**(2):545-553

[49] Reed DJ. Patterns of sediment deposition in subsiding coastal salt marshes, Terrebonne Bay, Louisiana: The role of winter storms. Estuaries. 1989;**12**(4):222-227

[50] Victor S, Golbuu Y, Wolanski E, Richmond RH. Fine sediment trapping in two mangrove-fringed estuaries exposed to contrasting land-use intensity, Palau, Micronesia. Wetlands Ecology and Management. 2004;**12**(4):277-283

[51] Allison M, Kepple E. Modern sediment supply to the lower delta plain of the Ganges-Brahmaputra River in Bangladesh. Geo-Marine Letters. 2001;**21**(2):66-74

[52] Goodbred SL Jr, Kuehl SA. Floodplain processes in the Bengal Basin and the storage of Ganges–Brahmaputra river sediment: An accretion study using 137Cs and 210Pb geochronology. Sedimentary Geology. 1998;**121**(3-4):239-258

[53] Van Santen P, Augustinus PGEF, Janssen-Stelder BM, Quartel S, Tri NH. Sedimentation in an estuarine mangrove system. Journal of Asian Earth Sciences. 2007;**29**(4):566-575

[54] Dubois RN. Seasonal variation of mid-foreshore sediments at a Delaware beach. Sedimentary Geology. 1989;**61**(1-2):37-47

[55] Thom BG. Mangrove ecology: A geomorphological perspective. In: Mangrove Ecosystems in Australia, Structure, Function and Management. 1982. pp. 3-17

[56] Untawale AG, Jagtap TG. Floristic composition of the deltaic regions

of India. Food and Agriculture Organisations of the United Nations; 1991

[57] Lugo AE, Snedaker SC. The ecology of mangroves. Annual Review of Ecology and Systematics. 1974;**5**(1):39-64

[58] Cahoon DR, Lynch JC. Vertical accretion and shallow subsidence in a mangrove forest of southwestern Florida, USA. Mangroves and Salt Marshes. 1997;**1**(3):173-186

[59] Capo S, Sottolichio A, Brenon I, Castaing P, Ferry L. Morphology, hydrography and sediment dynamics in a mangrove estuary: The Konkoure Estuary, Guinea. Marine Geology. 2006;**230**(3-4):199-215

[60] Cahoon DR, Hensel PF, Spencer T, Reed DJ, McKee KL, et al. Coastal wetland vulnerability to relative sea-level rise: Wetland elevation trends and process controls. In: Wetlands and Natural Resource Management. Berlin, Heidelberg: Springer; 2006. pp. 271-292

[61] Middleton BA, McKee KL. Degradation of mangrove tissues and implications for peat formation in Belizean island forests. Journal of Ecology. 2001;**89**(5):818-828

[62] Wolanski E. Transport of sediment in mangrove swamps. In: Asia-Pacific Symposium on Mangrove Ecosystems. Dordrecht: Springer; 1995. pp. 31-42

[63] Susilo A, Ridd PV, Thomas S. Comparison between tidally driven groundwater flow and flushing of animal burrows in tropical mangrove swamps. Wetlands Ecology and Management. 2005;**13**(4):377-388

[64] Susilo A, Ridd PV. The bulk hydraulic conductivity of mangrove soil perforated with animal burrows. Wetlands Ecology and Management. 2005;**13**(2):123-133

[65] De C. Mangrove Ichnology of the Bay of Bengal Coast, Eastern India. USA: Springer; 2019

[66] McKee KL. Biophysical controls on accretion and elevation change in Caribbean mangrove ecosystems. Estuarine, Coastal and Shelf Science. 2011;**91**(4):475-483

[67] Nyman JA, Walters RJ, Delaune RD, Patrick WH Jr. Marsh vertical accretion via vegetative growth. Estuarine, Coastal and Shelf Science. 2006;**69**(3-4):370-380

[68] Langley JA, McKee KL, Cahoon DR, Cherry JA, Megonigal JP. Elevated CO_2 stimulates marsh elevation gain, counterbalancing sea-level rise. Proceedings of the National Academy of Sciences. 2009;**106**(15):6182-6186

[69] Neubauer SC. Ecosystem responses of a tidal freshwater marsh experiencing saltwater intrusion and altered hydrology. Estuaries and Coasts. 2013;**36**(3):491-507

[70] Cahoon DR, Turner RE. Accretion and canal impacts in a rapidly subsiding wetland II. Feldspar marker horizon technique. Estuaries. 1989;**12**(4):260-268

[71] Pugh D. Changing Sea-Levels: Effects of Tides, Weather and Climate. Cambridge, UK: Cambridge University Press; 2004

[72] Cahoon DR, Reed DJ, Day JW Jr. Estimating shallow subsidence in microtidal salt marshes of the south-eastern United States: Kaye and Barghoorn revisited. Marine Geology. 1995;**128**(1-2):1-9

[73] Nolte S, Koppenaal EC, Esselink P, Dijkema KS, Schuerch M, De Groot AV, et al. Measuring sedimentation in tidal marshes: A review on methods and their applicability in biogeomorphological

studies. Journal of Coastal Conservation. 2013;**17**(3):301-325

[74] Phan LK, van Thiel de Vries JS, Stive MJ. Coastal mangrove squeeze in the Mekong Delta. Journal of Coastal Research. 2014;**31**(2):233-243

[75] Li X, Liu JP, Saito Y, Nguyen VL. Recent evolution of the Mekong Delta and the impacts of dams. Earth-Science Reviews. 2017;**175**:1-17

[76] Webb EL, Friess DA, Krauss KW, Cahoon DR, Guntenspergen GR, Phelps J. A global standard for monitoring coastal wetland vulnerability to accelerated sea-level rise. Nature Climate Change. 2013;**3**:458-465

[77] Lovelock CE, Bennion V, Grinham A, Cahoon DR. The role of surface and subsurface processes in keeping pace with sea-level rise in intertidal wetlands of Moreton Bay, Queensland, Australia. Ecosystems. 2011;**14**(5):745-757

[78] McKee KL, Cahoon DR, Feller IC. Caribbean mangroves adjust to rising sea-level through biotic controls on change in soil elevation. Global Ecology and Biogeography. 2007;**16**(5):545-556

[79] Cahoon DR, Lynch JC, Perez BC, Segura B, Holland RD, Stelly C, et al. High-precision measurements of wetland sediment elevation: II. The rod surface elevation table. Journal of Sedimentary Research. 2002;**72**:730-733

[80] Mudd SM, Howell SM, Morris JT. Impact of dynamic feedbacks between sedimentation, sea-level rise, and biomass production on near-surface marsh stratigraphy and carbon accumulation. Estuarine, Coastal and Shelf Science. 2009;**82**(3):377-389

[81] Banerjee K, Senthilkumar B, Purvaja R, Ramesh R. Sedimentation and trace metal distribution in selected locations of Sundarbans mangroves and Hooghly estuary, Northeast coast of India. Environmental Geochemistry and Health. 2012;**34**(1):27-42

[82] Chaudhuri P, Nath B, Birch G. Accumulation of trace metals in grey mangrove Avicennia marina fine nutritive roots: The role of rhizosphere processes. Marine Pollution Bulletin. 2014;**79**(1-2):284-292

[83] Watson EB. Changing elevation, accretion, and tidal marsh plant assemblages in a South San Francisco Bay tidal marsh. Estuaries. 2004;**27**(4):684-698

[84] Madsen AT, Murray AS. Optically stimulated luminescence dating of young sediments: A review. Geomorphology. 2009;**109**(1-2):3-16

[85] Krauss KW, Cahoon DR, Allen JA, Ewel KC, Lynch JC, Cormier N. Surface elevation change and susceptibility of different mangrove zones to sea-level rise on Pacific high islands of Micronesia. Ecosystems. 2010;**13**(1):129-143

[86] Horstman EM, Dohmen-Janssen CM, Bouma TJ, Hulscher SJ. Tidal-scale flow routing and sedimentation in mangrove forests: Combining field data and numerical modelling. Geomorphology. 2015;**228**:244-262

[87] Lesser GR, Roelvink JV, Van Kester JATM, Stelling GS. Development and validation of a three-dimensional morphological model. Coastal Engineering. 2004;**51**(8-9):883-915

[88] Hu K, Ding P, Wang Z, Yang S. A 2D/3D hydrodynamic and sediment transport model for the Yangtze Estuary, China. Journal of Marine Systems. 2009;**77**(1-2):114-136

[89] Adame MF, Neil D, Wright SF, Lovelock CE. Sedimentation within and among mangrove forests along a gradient of geomorphological settings.

Estuarine, Coastal and Shelf Science. 2010;**86**(1):21-30

[90] Ellison J. Holocene palynology and sea-level change in two estuaries in Southern Irian Jaya. Palaeogeography, Palaeoclimatology, Palaeoecology. 2005;**220**(3-4):291-309

[91] Wolanski E, Gibbs RJ, Spagnol S, King B, Brunskill G. Inorganic sediment budget in the mangrove-fringed Fly River delta, Papua New Guinea. Mangroves and Salt Marshes. 1998;**2**(2):85-98

[92] Smoak JM, Patchineelam SR. Sediment mixing and accumulation in a mangrove ecosystem: Evidence from 210Pb, 234Th and 7Be. Mangroves and Salt Marshes. 1999;**3**(1):17-27

[93] Thampanya U, Vermaat JE, Sinsakul S, Panapitukkul N. Coastal erosion and mangrove progradation of Southern Thailand. Estuarine, Coastal and Shelf Science. 2006;**68**(1-2):75-85

[94] Ha TTT, van Dijk H, Bush SR. Mangrove conservation or shrimp farmer's livelihood? The devolution of forest management and benefit sharing in the Mekong Delta, Vietnam. Ocean and Coastal Management. 2012;**69**:185-193

[95] Friess DA, Phelps J, Leong RC, Lee WK, Wee AKS, Sivasothi N, et al. Mandai mangrove, Singapore: Lessons for the conservation of Southeast Asia's mangroves. Raffles Bulletin of Zoology (SUPPL. 25). 2012:55-65

[96] Alongi DM. The impact of climate change on mangrove forests. Current Climate Change Reports. 2015;**1**(1):30-39